WORKSHOPS
master your skills

Easy Electronics

How To Use a Breadboard

Capturing Your Projects

More coming soon...!

brought to you by **Make:**

CONTENTS

R.U. Still a CYBERPUNK?

24

ON THE COVER:
Melissa Lamoreaux (makershare.
com/portfolio/melissa-lamoreaux)
is a maker who still lives the
cyberpunk aesthetic daily.
Originally from Los Angeles,
she now makes her home in the
San Francisco Bay Area where
she teaches programming and
robotics to children. In her free
time she creates technology to
care for plants and helps her
friends solve technical challenges.
She has previously modeled
Anouk Wipprecht's 3D printed
dresses.

Photo: Hep Svadja
Model: Melissa Lamoreaux
(dirtgeist.com)
PA: Pedram Navid and Jun Shéna
Styling: Brian Ford (insta: @BT4D),
Amy Martin, and Jun Shéna
Location: American Steel Studios
Font: Juan Hodgson

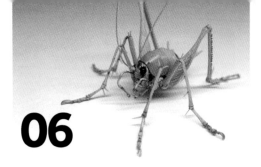

06

12

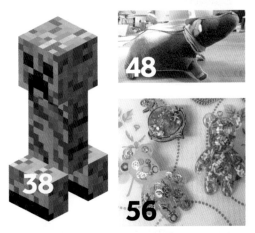

38

48

56

62

70

Hep Svadja, Juan Hodgson, Noriyuki Saitoh, courtesy of Les Machines, Mojang, Ashley Qian, Rachel "Konichiwakitty" Wong, Tim Deagan

Make:

> "The street finds its own uses for things."
> — William Gibson, "Burning Chrome"

EXECUTIVE
CHAIRMAN & CEO
Dale Dougherty
dale@makermedia.com

CFO & COO
Todd Sotkiewicz
todd@makermedia.com

EDITORIAL

EDITORIAL DIRECTOR
Roger Stewart
roger@makermedia.com

EXECUTIVE EDITOR
Mike Senese
mike@makermedia.com

SENIOR EDITORS
Keith Hammond
khammond@makermedia.com
Caleb Kraft
caleb@makermedia.com

BOOKS EDITOR
Patrick Di Justo

EDITOR
Laurie Barton

PRODUCTION MANAGER
Craig Couden

EDITORIAL INTERN
Jordan Ramée

CONTRIBUTING EDITORS
William Gurstelle
Charles Platt
Matt Stultz

CONTRIBUTING WRITERS
John Baichtal, Peter Bullen,
Erica Charbonneau, Tim
Deagan, Mari DeGrazia, Anne
Filson, Anna Kaziunas France,
Guillaume Grallet, Slater
Harrison, Kate Hartman,
Jarrod Hicks, Chris Hudak,
Bob Knetzger, Jean Perardel,
Ashley Qian, Jennifer Refat,
Gary Rohrbacher, Rick
Schertle, Becky Stern, Tshepo
Tshabalala, Rachel Wong,
Kitty Yeung

DESIGN,
PHOTOGRAPHY
& VIDEO

ART DIRECTOR
Juliann Brown

PHOTO EDITOR
Hep Svadja

SENIOR VIDEO PRODUCER
Tyler Winegarner

MAKEZINE.COM

ENGINEERING MANAGER
Jazmine Livingston

WEB/PRODUCT
DEVELOPMENT
David Beauchamp
Sarah Struck
Alicia Williams

CONTRIBUTING ARTISTS
21G, Vic Fieger, Juan
Hodgson, Bob Knetzger

ONLINE CONTRIBUTORS
Gareth Branwyn, Vivienne
Byrd, Chiara Cecchini, Jon
Christian, Jeremy Cook, DC
Denison, Stuart Deutsch,
Gretchen Giles, Liam Grace-
Flood, Brad Halsey, Ryan
Jenkins, Garrett Kinsman,
Ted Kinsman, Becky LeBret,
Sarah Vitak, Dr. Evan Malone,
Goli Mohammadi, Saba
Mundlay, Dan Royer, Dan
Scheiderman, Violet Su, Dan
Woods, Wayne Yoshida

PARTNERSHIPS
& ADVERTISING
makermedia.com/contact-
sales or
partnerships@
makezine.com

DIRECTOR OF
PARTNERSHIPS &
PROGRAMS
Katie D. Kunde

STRATEGIC
PARTNERSHIPS
Cecily Benzon
Brigitte Mullin

DIRECTOR OF MEDIA
OPERATIONS
Mara Lincoln

DIGITAL PRODUCT
STRATEGY

SENIOR DIRECTOR,
CONSUMER EXPERIENCE
Clair Whitmer

MAKER FAIRE

MANAGING DIRECTOR
Sabrina Merlo

MAKER SHARE

DIGITAL COMMUNITY
PRODUCT MANAGER
Matthew A. Dalton

COMMERCE

PRODUCT MARKETING
MANAGER
Ian Wang

OPERATIONS MANAGER
Rob Bullington

PUBLISHED BY

MAKER MEDIA, INC.
Dale Dougherty

Copyright © 2018
Maker Media, Inc. All rights
reserved. Reproduction
without permission is
prohibited.
Printed in the USA by
Schumann Printers, Inc.

Comments may be sent to:
editor@makezine.com

Visit us online:
makezine.com

Follow us:
🐦 @make @makerfaire
@makershed
google.com/+make
📘 makemagazine
makemagazine
▶ makemagazine
🎬 twitch.tv/make
📌 makemagazine

Manage your account online,
including change of address:
makezine.com/account
866-289-8847 toll-free
in U.S. and Canada
818-487-2037,
5 a.m.–5 p.m., PST
cs@readerservices.
makezine.com

WELCOME

Predicting the Future

BY MIKE SENESE,
executive editor of *Make:* magazine

YOU DON'T HAVE TO BE A TECHIE TO BE A MAKER, but there's a definite overlap, a fascination with new developments in tools and techniques of all types and an eagerness to put them to use in projects. For our staff, creating every issue of *Make:* includes the thrill of exploring the latest trends as the community finds and adopts them — and this issue even more so, as we focus on emerging technologies that should be on your radar, from computing to communications to crypto (see "Tech Trends," page 20). Predicting which research endeavors will turn into viable products is tricky, but we're up to the challenge.

Along those lines, there's never been a group that's as new-tech-forward as the cyberpunk community, which blazed its trails in the late '80s/early '90s and is again emerging in both aesthetic and concept. We visit this scene with a remake of the seminal poster "R.U. a Cyberpunk" from the influential tech magazine *Mondo 2000*, updated to reflect today's technology (page 24). From there we push into the realm of DIY spies and hackers, with a special section full of contraptions to build, components to salvage, and ideas to implement as you set off to become the latest 007. Have fun — but please don't break any laws!

ASK THE EDITORS

What's your favorite cyberpunk accessory and why?

Tyler Winegarner
Senior Video Producer
I remember being fascinated by monomolecular blades. Most were used to stab people, but I'd like to try a monomolecular band saw.

Hep Svadja
Photo Editor
A functional Power Glove-style armband computer. I want multiple OLED touchscreens, dedicated hardware buttons, and pop-up AR displays.

Roger Stewart
Editorial Director
Mirrorshades. So you can't tell what I'm thinking with my blank insectoid eyes reflecting your own face back at you — but now that I've told you why that sort of ruins it.

Patrick Di Justo
Books Editor
A small box that uses radio waves to send my thoughts to people around the world.

Matt Stultz
Digital Fabrication Editor
Ein the Data Dog from *Cowboy Bebop*. I liked him so much I got my own.

Issue No. 62, April/May 2018. *Make:* (ISSN 1556-2336) is published bimonthly by Maker Media, Inc. in the months of January, March, May, July, September, and November. Maker Media is located at 1700 Montgomery Street, Suite 240, San Francisco, CA 94111. SUBSCRIPTIONS: Send all subscription requests to *Make:*, P.O. Box 17046, North Hollywood, CA 91615-9588 or subscribe online at makezine.com/ offer or via phone at (866) 289-8847 (U.S. and Canada); all other countries call (818) 487-2037. Subscriptions are available for $34.99 for 1 year (6 issues) in the United States; in Canada: $39.99 USD; all other countries: $50.09 USD. Periodicals Postage Paid at San Francisco, CA, and at additional mailing offices. POSTMASTER: Send address changes to *Make:*, P.O. Box 17046, North Hollywood, CA 91615-9588. Canada Post Publications Mail Agreement Number 41129568. CANADA POSTMASTER: Send address changes to: Maker Media, PO Box 456, Niagara Falls, ON L2E 6V2

PRINTED WITH SOY INK

Budget Printers *and* Encouraging Educators

chiquimakers
ChiquiMakers

10 likes
chiquimakers Construyendo robots de papel 🤖😊🙌🤖👦 #Makers #makey @makemagazine #SomosMakers #ILoveRobotics #bucaramanga #colombia #VacacionesRecreativas
4 HOURS AGO · SEE TRANSLATION

SCHOOL LIBRARY OF THE FUTURE

I am a middle school librarian in Fredericksburg, Virginia and have been running a school library makerspace for more than four years. I wanted to start by thanking *Make:* magazine for all of the amazing resources that we have been able to utilize. Our program, Make It... Awesome!, is an affiliate member of Maker Camp, building the projects and having our students and their work featured in hangouts. Our students and I have also participated in numerous Maker Faires in our area, sharing their work. We are subscribers to *Make:* magazine and use this and lots of other *Make:* resources in our projects. *Make:* has also given me so much support as an educator to be able to share with my students. This encouragement and inspiration to make our library more centered on students' creativity has in part caused me to win the Virginia Association of School Librarians' Librarian of the Year Award, and I have been recently interviewed on ABC in Washington, D.C. for a segment about my award and library (makezine.com/go/sekinger).

–Nathan Sekinger, Fredericksburg, VA

NO LOVE FOR BUDGET PRINTERS?

In your 2018 desktop manufacturing shootout, I couldn't help but notice that the 3D printers start at $500. I'd be curious to know if that's because you can't get review models of the budget printers that are out there or if it's because you'd shy away from printers that perhaps aren't 100% ready to print out of the box or if there's another reason I'm not privy to.

–Rob Paige, Madison, WI

Digital Fabrication Editor Matt Stultz responds:

Hey Rob, thanks for reaching out. You may have missed the Monoprice Select Mini in the list, which definitely fits the budget printer category at $220, and earned our "Best Value" badge. I also think the Prusa i3 is a best-of-both-worlds printer: it's great out of the box but has a huge community behind it, making it even better with mods and hacks. It's received our top marks the last two years, and the MK2S kit is only $599. We aim to have a range of machines in our reviews and we will work on getting more budget models next time. ✎

Chiqui Makers, Hep Svadja

MADE ON EARTH

Backyard builds from around the globe

Know a project that would be perfect for Made on Earth?
Let us know: *makezine.com/contribute*

EXQUISITE ENTOMOLOGY

TAKE64.WIXSITE.COM/MUSI

At first glance, **Noriyuki Saitoh**'s insect collection triggers an instinctual recoiling, as any close encounter with a stinging wasp or prickly legged grasshopper will do. Each creation is so detailed and realistic, it seems it must be an actual insect that has crawled its way into the collection of bamboo-based renditions that you are expecting. But taking a cautious, closer look, you'll notice these are all indeed handmade works.

The 50-year-old Japanese artist has been crafting wooden insects for the past 10 years. His bamboo bugs are all actual size — meaning, quite small, which highlights his talent and creativity working with the medium. Delicate appendages link together with realistic joints; wings show accurate venation and membranes; and wispy antennae taper off into nothing. The critters come as standalone models you might see perched motionless on a branch, and in action poses: hornets building a nest, leafcutter ants carrying away foliage, a praying mantis ready to devour a just-captured butterfly meal.

To create his insects, Saitoh uses a standard assortment of hobbyist implements like X-Acto knives, precision tweezers, soldering iron, and so on. And while his collection of insects is extensive, there's one he likes best: "The most fun to make is the mantis," he says.

— *Mike Senese*

Noriyuki Saitoh

RISING FROM THE ROCK

MCCALLISTERSCULPTURE.COM

The arid heat of Arizona has become home to several mythical beasts and divine spirits. **Ryan McCallister** crea these sculptures by welding pieces c metal together into an exoskeleton a filling them with river rocks.

"I was inspired by the caged river rock walls, known as gabions, located around Phoenix and the Scottsdale area. They're either used for landscaping or the walls of a public bathroom. I thought they were a great medium, but felt like no one was pushing that concept to the limit McCallister says.

Operating out of Paradise Valley, McCallister creates and sells handmade décor, sculptures, and metalwork pieces. His first river rock sculpture was of Jörmungandr, a sea serpent from Norse mythology and archenemy of Thor, but he's since constructed the Minotaur, a griffin, ar several other creatures.

"With each new project, I find myse problem solving to achieve what's in my head, McCallister says. "It makes each piece of work a learning process The most difficult part of that process is certainly bringing my work to the next level. I want to capture people's imagination and bring attention to the craftsmanship, while also offering some element to help set me apart from other metal workers."

— *Jordan Rame*

Ryan McCallister

Gabriel Schama

CURVING CUTS

GABRIELSCHAMA.COM

Gabriel Schama lives and works out of Oakland, California, carving elaborate designs with his laser cutter that he's adorably nicknamed Elsie. "I studied art and architecture in college, but art has been a hobby, talent, and obsession of mine since I was a child," Schama says. "I worked in architectural metal fabrication and carpentry at various points before I struck out on my own as an artist," he continues, "So I acquired a lot of skills that are relevant to what I do now, but I didn't directly use a laser cutter until I bought my own."

Though he does draw, Schama relies on digital sketches for his professional work to achieve perfect symmetry without losing too many work hours. "On the production side, when I'm in full swing, I can crank out one to three pieces a day, depending on the size," he says.

Though Schama doesn't have any exhibitions coming up, he is participating in an interesting exchange with a tattoo artist. "I have a treasured relationship with Roxx of Two Spirit Tattoo," he says. "I've basically given her total freedom to do what she wants on my skin, and, in exchange, I'm adorning the walls of her studio with a series of large-scale laser cuts entirely in white."
— *Jordan Ramée*

TIME TO RIDE

USERS.HUBWEST.COM/HUBERT/CLOCK/CLOCK.HTML

When his son-in-law mentioned that his bike shop was looking for an art project, **Hubert van Hecke** set forth with an inventive twist on a timepiece.

The result is a frameless bike-parts clock that tirelessly spins away, reporting the time with both accuracy and flair. It uses weights, as opposed to a frame, to keep the chain taut as the whole mechanism sways to the passing of time. It's mesmerizing to watch.

Hecke waves his talents off as simple woodworking and metalworking skills that he was able to teach himself. "I've tinkered with clocks since I was little," he says. "I had the idea for this particular type of clock, with no frame and gears held by chains and gravity, many years ago."

He decided to make it bigger, and construct it with the pieces of a bike. "Hard to say how long the project took," Hecke says, "I worked on this on weekends when I had time, spread over a few months. Probably five weekends worth of time in total."

Hecke remarks that most of his projects follow the same build pattern: "Usually I make back-of-the-envelope sketches, and the details get solved on the fly."
— *Jordan Ramée*

Hubert van Hecke

MECHANICAL Masterminds

The awe-inspiring
Les Machines de l'Ile
are tackling a new addition
— a towering tree that
houses incredible surprises

Written by Guillaume Grallet

Innovation and technology reporter for
Le Point, **GUILLAUME GRALLET** has been
traveling in America, Asia, and Africa for
the last 20 years and has a huge interest in
the maker movement. He has interviewed
Elon Musk, Sergey Brin, Kira Radinsky, and
Rebecca Enonchong, among others.

Ajari

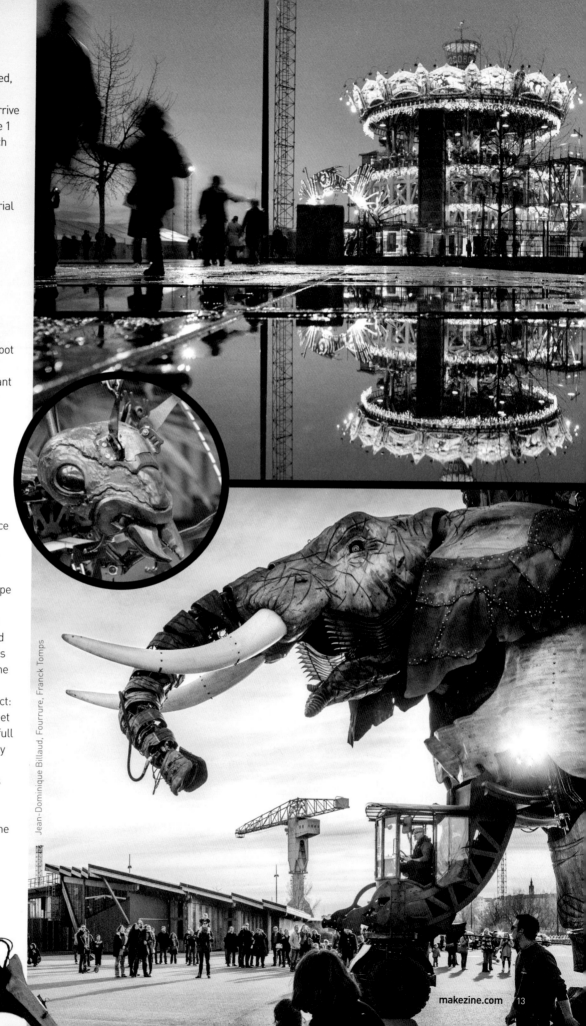

If you want to be astonished, go to Nantes in western France. As soon as you arrive in the city, take tramway Line 1 for a few stops until you reach the station *Chantiers Navals*. Cross the river and you will discover an incredible new world, a mix between industrial art and surprising creations inspired by Jules Verne, Leonardo da Vinci, and the British illustrator Heath Robinson. Welcome to *Les Machines de l'Île de Nantes*.

Inside the *Galerie des Machines*, their warehouse-sized stable if you will, a 40-foot dragon looms ominously. A four-story-tall, 47-ton elephant uses its trunk to spray water onto visitors as it carries up to 50 people on multiple levels on its back. There are humongous mechanical spiders and other fantastic creations which you can ride as they transport you into a unique and exciting experience throughout their plaza.

Overhead you will discover a branch of their Heron Tree (*L'Arbre aux Hérons*), a prototype created during the very beginning of Les Machines in 2007 to verify the strength and safety of the construction. This branch, stretching out from the Galerie and over the plaza, is the preface to their next project: a steel structure that is 164 feet in diameter and 100 feet tall, full of interactive delights that they aim to complete by 2022.

Pierre Orefice and François Delaroziere are, as they put it, the two co-authors of Les Machines. Pierre used to be the

Jean-Dominique Billaud, Fourrure, Franck Tomps

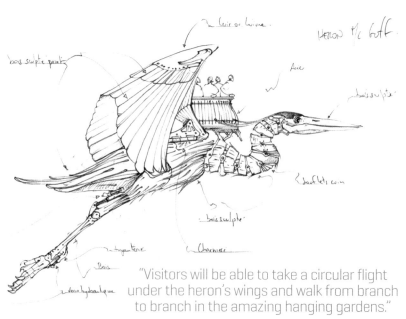

"Visitors will be able to take a circular flight under the heron's wings and walk from branch to branch in the amazing hanging gardens."

artistic director of the Manaus association, which created shows in urban spaces. He also co-ran the project Cargo 92, a trip on a boat with Royal de Luxe and the band Mano Negra that made stops throughout Latin America. "We had Christopher Columbus in mind and had the same exploration spirit," Pierre explains.

François, born in Marseille, has been working on very diverse projects with his company La Machine, like *Long Ma*, the mechanical dragon-horse created in 2014 for the 50th anniversary of the establishment of Franco-Chinese relations. (La Machine began as a precursor to Les Machines, and is still their design/build team.) He also has decorated *Le Dîner des Petites Mécaniques* — a restaurant where you are served by machines — and invented *La Princesse*, a 50-foot spider that debuted in Liverpool in 2008. In Nantes, he has created Les Machine's astonishing mechanical bestiary.

Inside Les Machine's glass offices, there's a reproduction airplane from the late 1900s. "We wanted to pay tribute to all the adventurers who were building the first machines to fly. More than half of them lost their lives in the process," Pierre explains. Close by, there is a small jungle where you can meet carnivorous plants. And then, not too far away from the warehouse is the *Carrousel des Mondes Marins* — an imposing three-level structure that can accommodate up to 300 people alongside sculptures of disturbing sea creatures

GROUP ENDEAVOR

The construction of *L'Arbre des Hérons* means the rebirth of the machines and jobs that were supposed to have disappeared a long time ago. More than 20 different jobs converge from this place of creation: designers, painters, molders, boilermaker-welders, cabinetmakers, sculptors, carpenters, cobblers, computer scientists, electric technicians — this undertaking will require the participation of at least 200 people.

"We act as if we are composing music," Francois says, explaining that he is the first step in a project's creation, and then his team joins in to contribute to its development. "Everybody will bring know-how to the project. Our machines will come alive!" he says.

The expertise of Yves Rollot, automation specialist, is extremely important. This scientist, who got a Ph.D. in robotics at Université Pierre et Marie Curie, will take care of the automation part of *L'Arbre*. The sculpting will be done by David Salviero and Emmanuel Bourgeau.

Guislaine Deguerry is in charge of color, while Bertrand Bidet takes care of the machine hydraulics during the show. Elodie Linard, ICAM engineer, coordinates all the work.

"The main area of research that we are now developing focuses on movement and expression in public spaces," François says. "We try to imagine the cities of tomorrow."

Rod Brazier Jean-Dominique Billaud, other photos courtesy of Les Machines

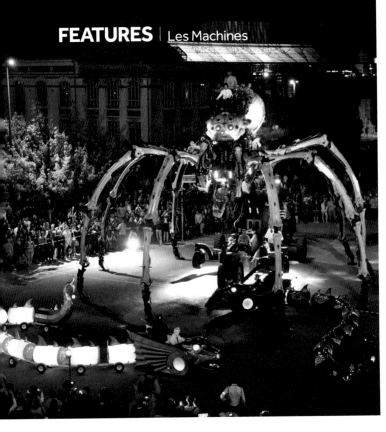

"The main area of research that we are now developing focuses on movement and expression in public spaces," François says. "We try to imagine the cities of tomorrow."

from all the world's oceans. Launched in 2012, its multiple levels each house different interactive elements including mechanical sea animals you can ride on and a stormy rotating scene where you can control an anglerfish puppet.

There is also a large illustration explaining the project of *L'Arbre aux Hérons*. Its 2022 debut in the park will be a new step for *Bas Chantenay* — a former granite quarry that is now a unique natural amphitheater. The gigantic tree — costing 35 million euros (which Les Machines plans to Kickstart in part) — will be able to welcome 450 people throughout its 22 zones. It will include suspended gardens and mechanical insects, while a double spiral staircase located inside the trunk will give visitors access to the different levels of its branches.

And at the top of the 1,000-ton steel tree will be two 50-foot mechanical herons that visitors will be able to climb on and take a flight overhead. The herons will carry 23 passengers for five minutes 150 feet above the ground, with views over the island of Nantes and the Loire river.

Why herons? "I wanted to work on something that represents Nantes. Visitors will be able to take a circular flight under the heron's wings and walk from branch to branch in the amazing hanging gardens," explains François. *L'Arbre aux Hérons* will make this part of the Chantenay quarry in Nantes one of the most beautiful urban gardens in the world. ◉

A NOTE FROM PIERRE AND FRANÇOIS

Dear Makers,
The Heron Tree project is an incredible artistic, technological, and industrial gamble. We cannot keep this adventure to ourselves. We need to share it, and share it with you!

Kickstarter is a formidable tool, here in Nantes but also worldwide, that will allow us to update you on every step of this 4-year project, from François' new sketches to photos of the 22 branches being assembled in the Chantenay Quarry. The more we share this project, the more powerful it will be.

The budget for the Heron Tree is 35 million euros. A third of it is reserved for private companies, but we need your support — and have wonderful rewards available in exchange. Our Kickstarter campaign runs from March 6 to April 17th.

Thank you in advance,
François Delaroziere et Pierre Orefice

Photos courtesy of Les Machines

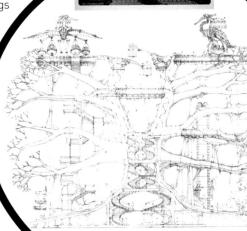

The Heron Tree concept sketch

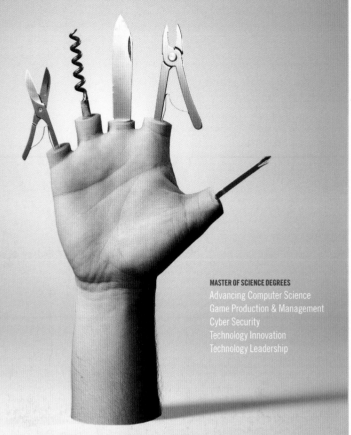

Open source toolkit **RootIO** is helping communities create their own micro stations

Rural Radio

Written by Tshepo Tshabalala

TSHEPO TSHABALALA is a web editor for journalism.co.za, JAMLAB Africa, and an M.A. Journalism candidate at Stellenbosh University. He holds an M.A. Philosophy, Politics and Economics and has worked for Thomson Reuters, *Forbes Africa*, and *Business Day* in South Africa.

The Journalism and Media Lab Newsletter (jamlab.africa) is a project of Wits Journalism and the Tshimologong Digital Innovation Precinct in partnership with Ryerson University and Journalists for Human Rights. **JAMLAB** supports innovators to bring new information, new ideas and new conversations to new audiences in Africa.

Radio continues to be a powerful medium across most of the African continent. Not only is radio used to share community information, but it is cheap and very accessible.

In Uganda, a mixing of radio's power with new mobile and internet technologies has created a cheap and powerful open source toolkit that allows communities to create their own micro-radio stations. All one needs is an inexpensive smartphone and a transmitter, and a community that will share, promote, and collaborate on dynamic content.

CONNECTING COMMUNITIES

RootIO is working to mobilize what they call "intercommunity communication." Co-founder Chris Csikszentmihalyi says the idea came after the earthquake in Haiti in 2010, when FM radio stations transformed their programming from ordinary radio shows to information on how people devastated by the earthquake could find water, or where they could find help. About a year and a half later, Csikszentmihalyi was in Uganda while working with the United Nations Children's Fund (UNICEF) through an educational program. He was amazed at the manner in which Ugandans used phones — rarely for calls.

"In the rural areas, people would go for long without recharging their credit or they didn't keep credit on their phone and at the same time, they listened to radio 24/7," he says. "In the villages where I was staying, people would walk about 7km to charge their phones, put credit on, and then only make a call. It's not like an always-on thing. They used it when they needed to. I thought, 'Is there a way of joining these two things together in a way that no one had done before?'"

While working at UNICEF, he met Jude Mukundane, who at the time was working for Uganda Telecom helping to develop mobile phone-based birth registration for the Ugandan Government in conjunction with UNICEF. Mukundane was doing some interesting stuff with telephones using Unstructured Supplementary Service Data (USSD). "I tried to hired him ... then about a year later, he was ready to do something," says Csikszentmihalyi.

"So together we said we should change radio and make radio work better with phones, make interaction with radio

RootIO tower in Uganda.

RootIO radio buckets.

easier for people. And we came up with RootIO at that point," Csikszentmihalyi says.

Mukundane became the chief technical officer, while Csikszentmihalyi focused on the fundraising, among other responsibilities.

"I make sure that the technology is at par with what we are promising communities," Mukundane says.

ATYPICAL TECH
The technology he is referring to is what makes these radio stations unique. They have no studio and all the shows are done using the host's smartphone.

How does this work? Users can purchase most of the materials at local markets. A small transmitter is built into a waterproof bucket with a fan, a charge-controller, and a smartphone, which is connected to an antenna and a solar panel.

The radio stations are really small and can serve a village or a couple of villages, reaching 10,000 listeners. The content produced by the radio hosts lives in the cloud so stations are able to share it.

"Our computer calls the station and the show's host," says Csikszentmihalyi. "The station's phone automatically answers when it is time for the programme. Listeners wanting to participate in the radio show's discussions would call in, but their calls would be dropped, then the computer would call them back. So people are not charged."

No one in the community is getting charged. RootIO buys data at bulk corporate rates that are about 50 times cheaper than anyone in the community could get. RootIO's costs are offset by selling advertising to NGOs and businesses.

BANDING TOGETHER
The team started with four stations two years ago and has now been commissioned to run another 12–15 stations in eastern Uganda by the Kenyan border, and five to seven radio stations commissioned in Cape Verde as well.

Csikszentmihalyi adds that they run RootIO at break-even. Even the software that they use is available free as open source on GitHub, and anyone can download and run the app.

Csikszentmihalyi and Mukundane hope to build a whole lot more inexpensive low-power FM radio stations, to hand control of FM radio to the people who depend on radio the most. ◓

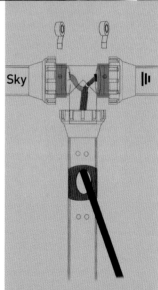

BUILD AN ANTENNA
Learn how to construct a low-power FM transmitter antenna from agricultural tubing at instructables. com/id/Low-Power-FM-Transmitter-Antenna-From-Agricultural.

TECH *Trends*

Keep an eye on these emerging developments and how you might incorporate them into projects

Written by *Make:* Editors

Lots of high technology is still too expensive or arcane for the day-to-day maker. But these days, theoretical research turns into commercial manufacturing faster than ever — and that means hot new tech is dropping into makers' laps. Here are some of the developments we're watching with optimism.

BATTERY POWER

Move over, AA batteries — **rechargeable 18650 Li-ion cells** have become a new standard. These can be used individually, outputting 4.2V, or wired together to form larger batteries. In fact, Tesla battery packs comprised thousands of 18650s until recently. **Tesla's new 2170 cell** (developed with Panasonic), promises even higher energy density.

Looking forward, researchers are testing **new materials** that may alleviate Li-ion's issues (flammability, limited raw materials). Aluminum-ion, dual carbon, and solid-state glass batteries are in the laboratory stage, and while they may never

materialize for production, they offer enticing theoretical advantages, including higher capacities and lower costs.

PROCESSING
Machine learning/artificial intelligence continues to gain attention, and more board manufacturers are optimizing products to allow for it. **TensorFlow**, Google's open source, machine-learning software package, is just over two years old and already utilized by many large companies. We're starting to see the maker community embrace it for projects that demonstrate intelligence, such as the $200 self-driving R/C cars being built by members of the DIY Robocars group.

For more powerful uses, **Nvidia** is leading AI development for both professional and high-end community groups with its latest Tegra chip-powered boards and Cuda platform. And Intel's new **Movidius NCS stick** is lowering barriers of entry for neural networking endeavors.

Beyond that, developments in **quantum computing** are quickly progressing — we'll have access to this on the community level sooner than most might expect. Last year, IBM gave Maker Faire attendees a crash course in the significance of quantum states and using quantum computing to solve everyday problems.

COMMUNICATIONS
Various new wireless communication protocols are emerging into the maker community. **Sigfox** and **LoRa** both offer wireless networks for low-power devices across wide areas, and are each incorporating their technology into prototyping boards from Arduino and elsewhere.

Optical communications are maturing as well. Manufacturers are starting to sell commercial devices offering **light-transmitted internet (Li-Fi)**, which presents a secure alternative to Wi-Fi. And **last-mile optical broadband** device makers like Koruza are opening new ways to build wireless

networks that rival fiber optics but without the infrastructure.

FABRICATION
The transition of **selective laser sintering (SLS)** from exclusively high-end industry to your desktop is underway. Formlabs Fuse 1 puts nylon SLS at the $10,000–$20,000 price point, which is in the realm of small design shops and midrange makerspaces. **Printing with metal** isn't far behind, with companies like Desktop Metal helping push down the price of using steel, copper, and high-tech alloys in digital fabrication.

There are also exciting advancements in digifab software — particularly, **generative design** and its ability to morph 3D models into lightweight, alien-looking structures that use less material but still offer the same strength characteristics as the original. Autodesk and SolidWorks are both building this into their software. Coupled with SLS technology, this will create a big shift for

prototyping and manufacturing in the near future.

SENSING
With many developments coming for sensors, we're keeping our eyes on **solid-state lidar**. Its increased robustness and lower prices will help transform the automotive field — both for autonomy and for standard driver safety — and will likely come out sometime this year.

CRYPTO
Some developers are dissecting the **blockchain technology** to offer new means of sharing content. At CES this year, Kodak announced its **KodakOne** rights management tool, which can allow photographers a new way to license and share their images and maintain their proper use through the terms of the agreement. Makers are also looking into this area with the announcement of the **Maker Token**, which advertises income potential for makerspaces and educators. ◔

Spies Like Us

The hackers, phreaks, and cyberpunks from the dawn of the internet can give us a glimpse into the future Written by Hep Svadja

CYBERPUNK — OUR COMMUNITY'S VERSION OF THE DIY SPY — exploded in the early 1990s with the advent of personal computing and on-the-go connectivity, firmly establishing its followers as digital rebels, instigators, and infiltrators both electronically and otherwise. The link between cyberpunk and making has always been strong, sharing the ethos "If you can't hack it, you don't own it" — along with a healthy interest in hacking the things other people might own as well.

This scene is now experiencing a resurgence, both in aesthetics and in its cultural viewpoint. As yesterday's cyberpunks have grown up to become today's maker futurists, it's worth looking to see what these heralds of the golden age of tech have to tell us about tomorrow. ✎

What is the coolest futuristic technology on your radar?

Mark Frauenfelder
Research Director at the Institute for the Future, former *Make:* EIC
"I'm interested in pseudo-haptic, which is a way to trick people into thinking they are feeling a physical change in something based solely on visual and audio information."

Gareth Branwyn
Futurism Journalist
"I am patiently waiting for self-driving cars to become commonplace. Because I have spinal arthritis and can't drive, a self-driving car would mean a huge improvement over my mobility and quality of life."

Jane Metcalfe
Co-Founder of *Wired*, Founder of Neo.life
"Hacking the body's lymphatic system when the liver fails. Using stem cells, you can convert the patient's lymph nodes to become like mini bioreactors that grow mini livers, which can start filtering blood wherever located."

Where do you see tech futurism going from here?

Cory Doctorow
Author, *Boing Boing* Editor
"I am immensely glad to say that I don't know because I am a strong believer that the future is contestable, not predictable: we get to make the future, we don't have to just accept what we get."

Julie Friedman Steele
Board Chair & CEO, World Future Society
"I see a return to human purpose, to backing away from technology for technology's sake and turning more towards technology for what we ultimately want to become as humans."

Limor Fried
CEO of Adafruit Industries
"The best way to see what the future holds is to make it happen yourself."

R.U. Still a CYBERPUNK?

Here's our modern update on the classic poster

Written by Chris Hudak

"DUDE — DID YOU KNOW YOU'RE ALL OVER THAT WALL, OVER THERE?!"

Now this is, objectively, *already* a spooky-enough string of words to hear slurred excitedly to/at you when you're rolling through an unfamiliar city on an ill-maintained shuttle-bus toward a hive-busy conference promising three straight days of next-stage tech *and* swarms of chemically augmented, socially stunted journalists from all over the globe.

I started up from my slump in the window-side seat, glanced blearily across the street — and damned if there I *wasn't,* staring square back at me over slightly-lowered mirrored sunglasses, from not one but *multiple* instanced-selves. His hair was considerably longer but there was otherwise no mistaking the prior-years me — the familiar leather jacket, the same chunky boots, the snarky get-a-load-of-*Moi* stare over the tipped mirrorshades ... and most of all, the scads of once-cutting-edge gear on my person and scattered around my feet. The guy I was looking at, and who was looking back at me, had been a writer (and in a few instances such as this one, a sort of poster-boy, I guess) for the seminal cyberpunk journal of all time: *Mondo 2000.* So, yeah; kinda spooky.

It is 2018 as I write this in Yokohama, Japan. The zombie *R.U. a Cyberpunk?* poster hadn't re-reared its '90s-rocker-haired head in years — after all, practically all the tech showcased in that original shoot had long since been obsolesced, miniaturized, or otherwise utterly eclipsed (quite a lot of that original trove's functionality and/or computing power could now fit comfortably in a 14-year-old's Hello Kitty phone case!). And then one day, I opened up a perfectly innocent-looking email from Stateside. And here we are. Which brings us to a word we need to address: *Cyberpunk.*

It's still a fine word — a pretty terrific, evocative word, in fact — and, yes, I would still deign to drag out that age-old, compressed-definition chestnut: "High-tech/Low-life." Don't let that latter half throw you; it's perfectly non-pejorative in nature. It means Attitude — and Ingenuity, of course — augmented by Tech, simple and pure as that. Cyberpunk is that tiny little cartoon mouse, the one on the mock-motivational poster titled *Defiance* — staring up from his inarguably "lowly" place at the huge, fearsome hawk screaming down at him, gleaming talons extended ... and giving said aerial predator the finger.

What you'll find showcased here is a completely updated list from that given in the original *R.U. a Cyberpunk?* poster. The street, as William Gibson has famously noted, finds its own uses for things; now it's your turn.

Hep Svadja, Juan Hodgson

1. Voice Changer: Back in the creaky-Net '90s, voice changers were readily available — they *sucked*, but they were available. Today, the spoofing is just that critical bit more difficult to spot, to actually warrant their occasional use.

2. Stun Gun: Maximally stunning.

3. ErgoDox Mechanical Split Keyboard with Custom Harness: Give new meaning to the words "Keyboard Cowboy" when you dual-wield this thing like pistols.

4. 3D Printed TSA Keys: A higher tech version of the Bic pen bike lock hack.

5. LED Glove: See what you're working with and look fly doing it.

6. VR Helmet: Snag the newest dev-rev now, because *this* is what is meant, or should be meant, by "*personal* computing." Sure, you'll be instantly hated, for a variety of reasons — but once upon a time you were seen in public with a Zune, so you can handle that.

7. Illegal-Megawatt Laser Pointer

8. Ceramic Switchknife: What can I say, I have a weakness for the classics; corrosion-resistant, superbly edge retaining, and nonmagnetic, because having to check your chrome at the door for a club full of amped-up death-metal knuckle-draggers is *bullshit*.

9. FLIR: Thermo-cool.

10. Raspberry Pirate Radio: As hacks go, it's a mulligan: a power source, the board itself, an SD card — and a single length of wire. The resultant broadcast-frequency, which can range between 1MHz and 250MHz, could well interfere with bands utilized by government and/or law-enforcement entities. (Not that *you* would be interested in that kind of thing.) makezine.com/go/pirate-radio-throwies

11. Bus Pirate: A hacker multi-tool that talks to electronic stuff. Probably the only thing you're interested in talking to anyway.

12. Cell Jammer: Jamming signals has its uses if you don't mind violating several laws in the process.

cyberpunk: ˈsībər ˌpəNGk/ n.
1: a late 20th to early 21st century techno-revolutionary, or someone who poses as such **2:** a hard-boiled hacker with anarchist inclinations **3:** a computer geek who likes the music of Perturbator **4:** a member of the counter-cultural "movement" of the same name, characterized by technological savvy and a rebellious lifestyle **5:** a long-anticipated game from CD Projekt Red studios **6:** someone who has delusions about living in the future **7:** someone who maintains that mirrorshades and all black everything never go out of fashion

The author in his original "R.U. a Cyberpunk?" article from Issue 10 of *Mondo 2000* magazine in 1993.

A. USB Battery Brick: Keep your device of choice charged every day for damn near a week.

B. FM Spy Transmitter: Whatever they're theoretically doing, the odds are *strongly* in your favor that they won't expect this half-a-century-old setup ... makezine.com/go/tiny-fm-spy-transmitter

C. JTAGulator: On-chip debug interface that can provide chip-level control of a target device to extract or modify data, on the fly.

D. DIY Raspberry Pi VPN/TOR Router: Anonymize your browsing from prying eyes with this hardware VPN build. makezine.com/go/pi-vpntor-router

E. Hideaway Voltmeter: Probably the single most innocuous item on this list, now rendered *utterly* suspicious by the mere fact you, dear reader, have stuck with this list. pokitmeter.com

F. Dremel Portable Soldering Iron: Butane-powered. Works as a flamethrower.

G. DSO Nano: Tiny single-channel oscilloscope fits in your pocket for signal analyzing on the go. Doubles as a vaporwave visualizer.

H. Dieselpunk Phone (Self-Build Burner): Obfuscating your available tech-level to prying eyes is a pretty cyberpunk notion — but you still won't get it past a metal detector. makezine.com/go/dieselpunk-phone

I. Pi-Top: Any hacker worth their bytes wouldn't be caught dead with off-the-shelf computing platforms.

J. MalDuino: You fancy yourself a "penetration tester," an appreciative dabbler in the cyber-transgressive arts; your ex-business partner or current section-head fancies you a nosy, potentially very dangerous, a-hole *extraordinaire*.

K. Macchina M2: Hack your car ... or someone else's. ✎

Be aware of what might be peeping on you

Written by Mike Senese

Jeepers Creepers

WITH THEIR SMALL SIZE AND UBIQUITOUS USE, WE'VE BECOME QUITE ACCUSTOMED TO COMMERCIAL HOME-MONITORING CAMERA SYSTEMS — so much so that they tend to fade into their settings, even when prominently placed up front and center. It's an extension of camera-equipped-everything maneuvering us to take the constant recording of our lives for granted.

That said, it's not OK to covertly record someone else without their permission or awareness. Yet, a quick search online reveals a chilling array of devices that are clearly designed for this type of endeavor — devices that make GoPros and babycams look as large and conspicuous as a flashing neon sign in Las Vegas.

Some discreet hidden camera gadgets that you can purchase:
- **Photo frame camera**
- **LED bulb camera**
- **Wristwatch camera**
- **Pen camera**
- **Eyeglasses camera**
- **Alarm clock camera**
- **Smoke detector camera**
- **Clothes hook camera**
- **Desk lamp camera**
- **Alarm clock camera**
- **Wall clock camera**
- **Charging dock camera**
- **Exit sign camera**
- **Cable modem camera**
- **Screw head camera**
- **USB charger and USB cable cameras**

You don't have to access the dark web to get these surveillance devices — they're all for sale on Amazon, most with Prime shipping. We live in weird times, indeed.

So what can you do to find these potential spying eyes? It can be a challenging task, but there are a few options.

LOOK FOR IR SIGNALS

Many cameras have an infrared LED bulb on them to illuminate a room outside of the visible light spectrum. These LEDs can emit a faint reddish glow in low- or no-light conditions — if you notice an odd glow, it might be coming from a hidden camera. A phone camera will also show an active IR LED glowing purple, as mentioned in "Magic Touch Frame" on page 33.

If you want to be more active in finding IR emissions, use your own IR-capable monitoring camera (buy a cheap USB version online) to scan the room to look for otherwise invisible shining sources. They should show up like a spotlight.

LISTEN FOR ELECTRONIC GLITCHING SOUNDS

Some cell phones will emit a crackling sound when exposed to electromagnetic signals like those that surveillance cameras can emit. Move your phone around the suspected area of a hidden electronic device to check for this telltale interference.

SEARCH FOR LENS REFLECTIONS

You can do this with smartphone apps or pricier hidden camera-detecting devices. The method works by having your phone's flash illuminate a space as you pan around; it looks for glints of light that can reflect off a camera's lens, even covertly placed ones.

SCAN FOR RF SIGNALS

Wireless cameras transmit their signal in some manner, often via radio waves. Dedicated camera-detection tools look for these signals to help pinpoint an apparatus' presence.

If you do encounter an illicit device, the experts suggest disabling it and notifying the authorities. Be safe! ◉

Hep Svadja, Juliann Brown, 21G, DHGate, Censee, Prweyn, MiniGadget, Kingnet

Hide and Seek

Make these clever devices and get your covert career underway

Written by *Make:* Editors

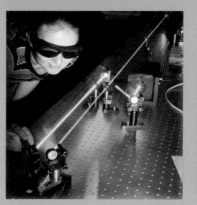

MAYBE YOU'VE ALWAYS DREAMED OF BEING A CIA OPERATIVE. Or maybe you're really into 007. Whatever the reason, you've always had a thing for spies and those high-tech toys they have up their sleeves.

Well, now you don't need all that formal training, not to mention all that danger. You can be the spy you've always dreamed of being with a little help from the DIY crowd and a little savvy. Here are 8 cool and geeky gadgets you can create yourself to impress your friends with your spy powers:

1. SECRET CHESSBOARD COMPARTMENT

Build this sneaky chessboard and you'll have a secret storage spot no one will ever suspect. Moving the chess pieces across the board in a particular way will unlock a secret compartment that you can use to hide diamonds, baseball cards, or the evidence linking you to the crime of the century. makezine.com/projects/secret-chessboard-compartment

2. HOLLOW BOLT "DEAD DROP" DEVICE

When spies need to pass off a note or microfilm without meeting in person, they sometimes use a "dead drop" device. An effective dead drop device looks common enough to blend in without causing suspicion under the casual glances of passersby. This hollowed-out bolt is one such item. makezine.com/projects/dead-drop-device

3. BUG-IN-A-BOOK

Eavesdrop on any devious plotting through an ordinary FM radio with this project. Connect a shirt-pocket "amplified listener" hearing aid with an in-car FM transmitter to create a wireless bug. Then tuck them inside a hollowed-out book with the microphone concealed by the dust cover, and you'll have a covert listening device that you can leave lying around or on a shelf near a surveillance target. makezine.com/projects/bug-in-a-book

4. PLANT A BUG

Another easy way to listen in on any conversation, anywhere. All you need is a cellphone and a plug-in headset (wired, not Bluetooth).

Connect the headset. On the cellphone, turn off both the ringer and the vibrator and turn on auto answer. Hide the phone and headset in the location you want to bug. You're ready to go. Now all you have to do is call the cellphone, which will answer after a couple rings and you can secretly hear everything that's being said. makezine.com/go/plant-a-bug

5. COFFEE CUP SPY CAM

Modify an everyday coffee cup to house a concealed camera that snaps a picture every time you take a sip.

The trick is to use two paper coffee cups — install the device in one, slide it into the second, and align holes cut in the bottoms of each. Two LEDs shine through the plastic lid — one illuminates when the tilt switch is activated, the other flashes twice after a picture has been taken. makezine.com/projects/coffee-cup-spy-cam

6. REMOTE ALARM

A tripwire is one of the most basic ways to set up a simple security system. You run a line across a pathway. Then when someone walks through the line, it activates an alarm. This kind of system is easy to deploy and is fairly effective. But there is always room for improvement.

The most inconvenient thing about a classic tripwire alarm is that it requires you to run a physical line from the tripwire to the alarm. This makes it difficult to set up a system where the alarm is far away or inside a building. To get around this problem, this simple remote system uses a small radio transmitter to activate the alarm wirelessly. makezine.com/projects/remote-tripwire-alarm

7. LASER TRIPWIRE

No security system is complete without lasers. This project shows you how to build a laser tripwire alarm from a laser pointer, a couple of mirrors, and a few dollars of electrical parts. With this you can cover an entire house with an array of light beams. If any one of them is crossed it sets off your alarm. It can be a standalone alarm or it can be integrated into a larger DIY security system. makezine.com/projects/laser-tripwire-alarm

8. SECRET MESSAGE ON THE BOTTOM OF A GLASS

Fans of Patrick McGoohan's classic BBC spy series *The Prisoner* will recognize this gimmick from Episode 15, "The Girl Who Was Death." This is a pint glass with words etched on the bottom (in authentic "Village" font) that appear line by line as the liquid is imbibed. makezine.com/projects/you-have-just-been-poisoned ◗

Garry McLeod, Tyler Winegarner, David Simpson, Hep Svadja, Jason Poel Smith, lightpoet - Adobe Stock, Sean Michael Ragan, Guy Sagi - Adobe Stock

Invisible Touch

Build a Magic Frame that turns any surface into a giant "touch area" interface!

Written by Jean Perardel

MAGIC FRAME

EVERYTHING BECOMES
A TOUCH AREA

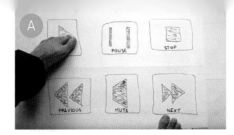

WHAT IF A SIMPLE FRAME COULD TRANSFORM ANYTHING INTO A "TOUCHSCREEN"? What would you do? Control music with your feet (Figure Ⓐ)? Turn your coffee table into a giant game controller? Slide your finger down the wall to dim your lights? You could even control your whole house secretly from a painting!

Here's how you can build your own touch controller for use on any surface. It uses invisible light triangulation technology to detect where your finger(s) are placed within the frame. This method is quite inexpensive and can be adapted to very large surfaces if you play around with LED power.

Of course, nothing is perfect. With this solution it's hard to cover the entire frame area perfectly, so it's not possible to build a 100% accurate multi-touch surface. But it works, it gives all kinds of interesting results, and I'm sure there are a lot of possibilities for improving it.

HOW DOES IT WORK?

Light triangulation is a simple but powerful way to read coordinates. Basically, you're using a large number of light detectors with LEDs to shine on them. You turn on each LED, one by one, and read the sensors each time. When a sensor can't see any LEDs, it means that an object is blocking the light. With enough LEDs and sensors, you can derive a fairly precise position.

I developed 2 different solutions to build my DIY touchscreen:

IR SENSORS — This solution uses a large number of infrared (IR) LEDs and IR photodiodes facing each other (Figure Ⓑ). When an LED is lit, several photodiodes can receive the emitted light and "see" if an object obstructs the light or not. You could add LEDs and sensors on all 4 sides of the frame to maximize coverage and precision, but here we'll just use 2 sides as shown.

CIS SENSORS — The second solution is more complex (Figure Ⓒ). *Contact image sensors* are present in most flatbed scanners. In those devices, the CIS is basically a black and white camera that reads only one line of pixels. To scan a color document, an RGB LED blinks in three colors (red, green, and blue) and the CIS reads all of them for every line of the document. Then it can calculate the exact colors. Using a CIS sensor gives you 2,700

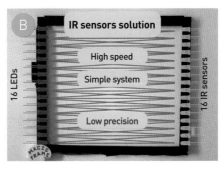

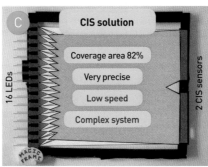

light sensors on a 20cm line, which creates great precision with just a few LEDs. But, CIS sensors are harder to find and to hack, so I'll show only the IR solution here.

The Magic Frame is built using modular 3D printed elements with IR LEDs and photodiodes installed in them. A variant includes room for an LED driver chip. These elements can be assembled to form an adaptive frame as big or small as you need.

HOW TO DRIVE SO MANY LEDS AND PHOTODIODES?

We need a huge number of digital pins to control the IR LEDs, and a lot of ADC pins (analog-digital converter) to read all those photodiodes. As you may know, on the Arduino Uno microcontroller board there are only 14 digital pins and 6 analog ... not enough.

So I used a Teensy 3.6 board. It's fully compatible with the Arduino environment and it fits better in this project for several reasons:

● It's way faster — 240MHz instead of 16MHz, 32-bit calculations instead of 8-bit

● It has a lot of inputs/outputs — 58 digital, 24 analog — in a very small package

TIME REQUIRED:
Build: 16–20 Hours
3D Printing: 30–70 Hours

DIFFICULTY:
Moderate

COST:
$60–$110

MATERIALS
FOR A 26CM × 26CM FRAME:
» **Teensy 3.6 microcontroller board** with USB cable
» **Z-Uno microcontroller board (optional)** if you want to control Z-Wave home automation
» **LED driver chip, STP16CP type** SMT package (the DIP package was discontinued)
» **SSOP-24 SMT breakout board**
» **Infrared (IR) LEDs (16)** 3mm, 20mA, 30° viewing angle
» **Infrared (IR) photodiodes (16)** same wavelength as your LEDs; typically 940nm
» **Frame** 3D print my modular frame (get the files for free at github.com/jeanot1314/Magic_Frame), or use your own frame.
» **Solid core wire, insulated, fine gauge** such as wire wrapping wire, 24AWG–30AWG
» **Headers, 2.54mm: male (36) and female (36)**
» **Capacitor, electrolytic, 10μF**
» **Resistors, 220Ω, ¼W (9)**
» **LiPo battery, 3.7V, 200mAh (optional)** if you add the Z-Uno

TOOLS
» **Soldering iron and solder**
» **Computer, with Arduino IDE and Teensyduino plugin** free from arduino.cc/downloads and pjrc.com/teensy/td_download.html
» **Project code** free from github.com/jeanot1314/Magic_Frame
» **3D printer (optional)** You can send the files out for printing the frame, or drill holes in an ordinary picture frame instead.
» **Wire cutters / strippers**
» **Multimeter (optional)** helpful for testing your soldering

Hep Svadja, Taras Livyy - Adobe Stock, Jean Perardel

JEAN PERARDEL
lives in Grenoble, France. Born in the Alps, he loves trekking, paragliding, and creating things. He has been playing with Arduino for 7 years now, and he still isn't tired of it.

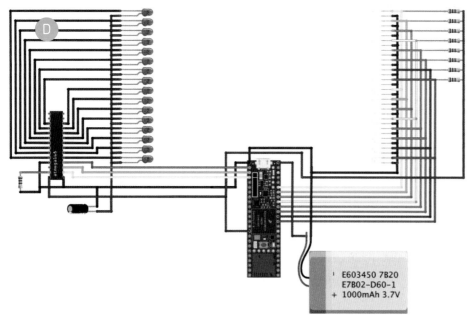

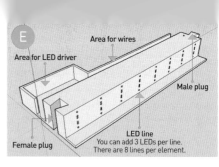

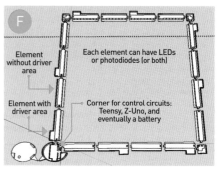

Jean Perardel

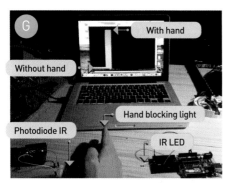

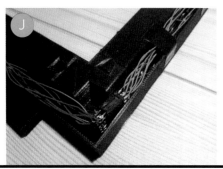

● Your computer can recognize the board as a USB input — like a keyboard, mouse, or joystick — which means you can use your Magic Frame touch surface as a remote control for all kinds of devices, or even a game controller, to play with a Raspberry Pi emulator, for example. Even though we now have enough digital pins, I added an LED driver board to make the build more flexible. The STP16CP type are common, inexpensive, and not too complicated to drive. You can drive 16 LEDs per chip, and connect as many chips as you want in series. Perfect for my modular frame!

For the analog pins, 24 still isn't that many for a big frame, so I used a little trick: I wired two photodiodes on the same ADC. As my photodiodes don't have a very large detection angle, one LED can only illuminate 3–5 of them, and I've wired them so there's no overlap. Because we light each LED separately, we know we're reading the 3 sensors facing that LED, not other sensors on the shared ADC.

So now you can see what's going on in the schematic (Figure D). For a 26cm × 26cm frame, you'll build one of the driver circuits shown, and one of the photodiode chains. Repeat these for a bigger frame. As you can see, the LED driver controls the cathodes of the LEDs; this way, the driver chip doesn't need to supply too much power. The LiPo battery is optional, for adding a Z-Wave board (see "Improvements" on the next page) or other wireless communications.

1. 3D PRINT THE FRAME

Start your build by printing 7 plain elements, 1 element with STP driver area, 3 corners, and 1 "control" corner (Figures E and F). It's a long job, as each element needs 4 to 8 hours to print, but you can speed things up by increasing the layer height; it will be less precise but faster to print. If you'd rather use an existing frame, you can drill holes for the IR LEDs and receivers, and mount the electronics on the back.

2. UNDERSTAND THE LED AND PHOTODIODE REACTIONS

Once you've got the electronics together, let's test our program. You can wire one LED and one photodiode and upload the AnalogReadSerial sample program from Arduino (File → Examples → 01.Basics → AnalogReadSerial) to read the coordinates (Figure G). I've also added a Graph program that you can open with Processing to view a nice graphical result.

Here's where you'll want to know the actual specifications of your LEDs and photodiodes. Some LEDs have wider angles, handle more power, or are brighter on the sides than in the middle. I used standard 3mm IR LEDs rated for 20mA power and a 30° angle. These specs work fine for a frame that's 2 elements wide (26cm, as each element is 13cm); the 30° angle is enough to light the 3 facing sensors. If you want to build a larger frame, you'll have to multiply the LEDs per line (3 maximum), or use higher-power LEDs like 50mA.

If you use LEDs with a wider viewing

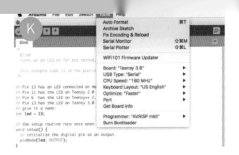

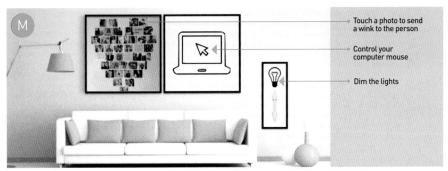

- Touch a photo to send a wink to the person
- Control your computer mouse
- Dim the lights

angle, you'll be able to light more sensors and increase the precision. But be careful, a bigger angle will make the light weaker on each sensor.

TIP: Here's a very useful trick. Do you know how to "see" an infrared LED? Turn on your phone's camera and look at the screen. While your eyes can't see IR light, the camera can, and retransmits it as a purple color, as you can see in Figure H.

3. START SOLDERING

This is another long part of the build … but you should soon have your Magic Frame ready. I strongly recommend you use very thin wire to solder the LEDs and photodiodes. It will make things much easier and take up less room in the frame (Figure I).

Another trick is to use headers to make it possible to adapt your frame to different sizes later (Figure J). These can be useful for debugging too.

4. TIME FOR PROGRAMMING

The Teensy board is fully compatible with the Arduino environment, you just have to install a little plug-in that you can find at pjrc.com/teensy/td_download.html.

Connect the Teensy to your computer with the USB cable, then select the Teensy from the Tools → Board menu (Figure K), and select your USB type (Serial, if you want your device to act as a mouse, keyboard, etc.). Download the project code from github.com/jeanot1314/Magic_Frame, open

it in Arduino, and push it to the Teensy using the Upload button. Now open the Serial Monitor window and you'll see readouts from your sensors. Verify that each sensor can see the 3 LEDs facing it. You can check by using your hand to block the light. Your Magic Frame is operational!

Now you can connect your Magic Frame to a USB port on any computer, and start experimenting.

5. FINAL ASSEMBLY

Figure L shows my final build of a large Magic Frame; you can, of course, adapt the size to your needs. I even 3D printed a logo for it.

USE IT

Let Your Fingers Do the Talking

So now you can transform any surface into a touch area, but what can you do with it? You can mount your Magic Frame on any surface — a tabletop, a fridge — but one fun approach is to frame images on the wall, and then map secret functions to those images (Figure M).

Touch a person's face to send them a "wink" online. Touch a favorite album cover to magically start playing tunes. Remotely control your computer's mouse, or scribble a code word with your finger to open a secret door. Anything you could do with a mouse or joystick, you can try with your Magic Frame.

Follow this project at makershare.com/projects/magic-frame-everything-touch-area and share your ideas. ⏻

IMPROVEMENTS

USE CIS SENSORS — As I mentioned earlier, you can also use a CIS sensor from a document scanner as your light detector. This method will improve your precision a lot, as the CIS has more than 2,700 light sensors on it. Figure N shows what they look like.

I also made a project based on this technology: a coffee table to play games. It won the 2016 Raspberry Pi contest on Instructables (Figure O). Learn more at youtu.be/Shyj-7JLFsg.

ADD Z-WAVE CONTROL — For home automation, you can add a Z-Uno board to control your Z-Wave peripherals, like smart plugs or lights. I used mine to control a Fibaro lamp dimmer.

Wire the Z-Uno to the Teensy as a UART serial peripheral; also add a ground as a reference for the boards, and a VCC so the Teensy will power the Z-Uno (Figure P). This way, the two boards can communicate and the Teensy can ask the Z-Uno to send specific messages.

The Z-Uno can be also programmed with the Arduino environment and a plugin. I added some Z-Uno code to control a smart plug to the Git for this project; you can see a little video about it at youtu.be/_KhWwn-HI6w.

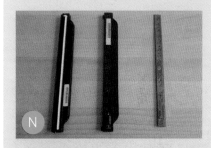

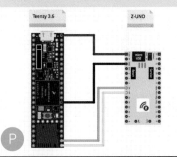

Urban Ore

Fill your parts bin with these valuable scraps, freely available to those in the know
Written by Caleb Kraft

A RESOURCEFUL MAKER IS CONSTANTLY ON THE LOOKOUT FOR CLEVER BUILDING MATERIALS. This can mean scanning the curbsides as you drive, browsing through the free section of Craigslist, and taking an occasional trip into the land of dumpster diving. When you see a pile that you just know would ordinarily cost you a considerable sum for the parts alone, the elation is unavoidable.

Just because something is free or cheap doesn't make it worthwhile. That said, cracking open these eight finds nearly always reveals a gold mine of parts inside.

1. OLD COMPUTERS

Computers are stuffed to the brim with parts that can be repurposed in other projects — it's kind of amazing that people give them away for free. Some of the most common components you can scrounge from a broken computer on the curb:
- Fully working power supplies
- Lasers!
- Big, fat, pretty heat sinks
- Stepper motors from the CD and floppy drives
- Matched rack-and-pinion sets from CD drives
- Strong magnets from the hard drives

Project: Turn a PC power supply into a bench power supply — makezine.com/projects/computer-power-supply-to-bench-power-supply-adapter

2. PRINTERS, SCANNERS, AND FAX MACHINES

There are times when it is literally cheaper to buy a new printer than it is to buy new ink for it. This is incredibly wasteful and results in tons of printers showing up for free on Craigslist or in dumpsters. If you're doing robotics though, printers and the like are sources for some great parts:
- DC motors
- Stepper motors
- Optical sensors
- Smooth rod
- Geared motors with matching belts

Project: Wind Lantern — makezine.com/projects/wind-lantern

3. PROJECTION TVS

These hulking dinosaurs are dying out due to the popularity of flat panel TVs — good for makers as they keep showing up for free. If you need some optics, these things can be a fantastic source:
- Massive Fresnel lens
- Smaller lenses for the individual projectors, housed within (usually 3)

Project: Giant Fresnel Solar Heat Ray — makezine.com/2011/06/25/solar-sinter-project-3d-printing-with-sunlight-and-sand

4. VCRS

VCRs may be antiquated now, but they're still available. And they have some pretty cool components:
- A hefty and solid rotary encoder (the VCR head)
- Linear actuators
- Springs
- DC motors
- A timing circuit

Project: VCR Cat Feeder — makezine.com/projects/vcr-cat-feeder

5. POWER WHEELS

Once the gas pedal stops working on these, they often end up in the dumpster. However, the drive motors and battery are often still perfectly fine. Those powerful motors may not be fast, but they've got decent torque for something you got for free. We've even repurposed their plastic gearboxes for simple locomotion projects.

Project: Power Racing Series Clown Car — makezine.com/projects/powerwheels-clown-car

6. RADIO CONTROLLED TOYS

Cheap R/C toys are everywhere now. This means that as the flimsy plastic bodies give out, the good stuff gets tossed with the rest. You can repurpose a cheap remote to trigger your Arduino, and use the chassis for your next rolling robot.

Project: Controlling an R/C toy with a USB steering wheel — makezine.com/2015/08/10/driving-an-rc-car-with-arduino-and-a-usb-racing-wheel

7. TIRES

Tires are a pain to get rid of. You often have to pay to have them hauled off. Many even feel relief when a maker knocks on their door asking if they can take away that pesky pile of old tires.

Little do they know, you're collecting the raw materials to use in an incredible sculpture, or possibly getting new soles for your homemade shoes. Get yourself a standard desktop metal shear and you can cut through them like butter.

Project: Running shoes with tire soles — makezine.com/craft/flashback_retro-style_running

8. OLD CLOTHES

If the Swap-O-Rama-Rama has taught us anything, it is that there is no such thing as trash fabric. Any fabric is good, and frankly, you can find some really nice material in piles of things being given away for free.

Project: Make a garden apron from upcycled jeans — makezine.com/craft/make-a-quick-garden-apron-from-upcycled-jeans ⊘

pavelkubarkov / Adobe Stock, Derek Tsang, Daniel Hoherd, Daniel Lobo, Hobvius Sudoneigm, LvL1 Hackerspace, Les Chatfield, Jayme del Rosario, Magnus D

Light Security

Perfect your evasion skills with this laser-armed maze
Written by Peter Bullen

Hep Svadja, Peter Bullen

PETER BULLEN
is an interdisciplinary maker interested in light, movement, music, and technology.

THE LASER MAZE IS THE ULTIMATE MOVIE-THEMED SECURITY SYSTEM, which protects valuable artifacts with a web of laser beams that would-be-thieves must stealthily duck, weave, crawl, and jump through without touching any beams and setting off the alarm. In real life, laser mazes don't make the best home security systems, but they are a lot of fun to build, and even more fun to play with.

> **CAUTION:** Laser light is intense and can damage your eyes. Remove highly reflective surfaces such as mirrors that may redirect laser light into someone's eyes unexpectedly. Be sure to only use laser powers of 5mW or less to reduce the chance of eye damage in case of accidental eye exposure.

THE BASIC LASER MAZE

Building a basic laser maze is fairly simple: You'll only need PVC pipe and connectors, "constant on switch" laser pointers, modeling clay, and a fog machine. Cut the PVC pipe into groups of equal length anywhere between about 2–3½ feet and connect them to form a large cage-like structure (Figure Ⓐ) that you can walk through (6½ feet high, 5 feet wide, and 10 feet long is a good start). Mount the laser pointers on the sides of the PVC pipe structure with modeling clay, direct them into the center space at different angles, and turn them on. Finally, turn the lights off and pulse on the fog machine to make the laser beams appear bright and beautiful.

For best results, go with the higher quality laser pointers with a "constant on switch" that are designed for extended use. Any color laser pointer works, but green has the brightest effect.

LASER DETECTION

A fun add-on to your laser maze is a detection system that sounds an alarm or flashes a light when someone blocks one of the beams. Mount a photoresistor on the opposite side of the structure from each laser pointer. Wire each photoresistor into a voltage divider circuit and connect it to one of the analog inputs of an Arduino board. When the laser shines on the photoresistor, its resistance is low, but as soon as the beam is blocked, the resistance jumps up, and the voltage to the Arduino pin increases. Program the Arduino to detect this voltage jump and trigger an alarm.

MOVING LASER BEAMS

A moving laser maze creates a whole new experience. Attach each laser pointer to a motor mounted on the PVC pipe structure (Figure Ⓑ). Program each motor to rotate back and forth at different speeds (5–10 seconds per cycle works best) and over various angles to make endless possibilities of changing laser obstacles. You may want to 3D print mounts and custom boxes to keep everything organized.

For detection, mount a large semi-transparent sheet opposite the laser pointers (Figures A and Ⓒ), so each laser makes a moving dot on the sheet. Set up a camera outside the maze so it can see the entire sheet. You can build a computer vision (CV)program to count the moving dots viewed by the camera, in order to detect when a laser has been blocked. For CV beginners, *scikit-image* is a good package to start with for Python, and OpenCV works with multiple programming languages. ◗

JOHN BAICHTAL has written over a dozen books, with topics ranging from Arduinos to Lego to robots. His current project is the *LED Project Handbook*, forthcoming from No Starch.

A

Craft a Creeper

Written by John Baichtal

Build a motorized "mob" robot and attached controller

When I wrote my new book *Make: Minecraft for Makers*, you know I had to include a monster Creeper project. Here's how you can build a motorized Creeper, with a metal skeleton and wooden skin. Aside from the fact that this thing most certainly doesn't blow up, you'll love it, and you'll learn a lot about robotics and Arduino along the way. Let's get to it!

The Creeper consists of a robot chassis kit with add-on parts creating the mob's distinctive armless body, with a servo motor to move the head around. Begin by taking a look at the Creeper in-game. Just be sure to stick to Creative mode or you may find yourself getting blown up!

The Creeper (Figure Ⓐ) has a cubical head 8 pixels on a side, a 4×8×12 body, and four 4×8×4 legs. It's actually a pretty elegant design, which makes it a breeze for building a physical re-creation.

DESIGN IT

The Robot Creeper seems super challenging at first. The thing has to look like a Creeper, ideally proportionate with the game element. At the same time, it also has to function as a robot. In other words, regardless of its outer appearance, the Creeper has to be able to fit all the necessary robotic components, particularly the chassis kit we're using for the base.

I began with the Actobotics Bogie Runt Rover, a kit available for around $70 that comes with a chassis, six motors, and six wheels. The assembled rover's chassis measures 6"×9", though the wheels project a little, and it rides fairly high: 6" off the ground. With those measurements I was able to decide the size of the footprint: 12"×8" — conveniently, one inch per pixel.

Applying the one-inch scale across the whole robot makes for a 12"-high, 8"-wide, and 4"-deep body, an 8" cubical head, as well as four legs 6"×4"×4". However, for the legs I decided to merge the front pair and back pair into 8"-wide blocks — the thing is going to roll, not walk. Figure **B** shows my final design. I created vector files to be used on a laser cutter, but you could simply cut the pieces from wood, or get creative with other materials — recycle those Amazon boxes and use some packing tape to knock out a cheap, simplified version.

Next, we need to design the robot's electronics. What will be its functionality? How will it be controlled? The Minecraft Creeper is known for blowing up, and clearly that was out. It also turns its head, and we can do that by putting a servo motor inside the body to turn our robot's head. The Creeper also has eyes that turn red when it's about to explode. That's easy! We'll put NeoPixel Jewels in the head.

Under the hood I stuck with the classic Arduino Uno, with a motor control shield sitting on top. This add-on board helps the Arduino manage the high voltages needed to run motors, and it simplifies controlling them.

Speaking of control, I'm making a basic controller (Figure **C**) that connects to the robot via a trio of wires.

BUILD YOUR CREEPER BOT
1. ASSEMBLE THE ROVER KIT
Begin the build by tackling the Bogie Runt Rover (Figure **D**) kit. There's a great assembly video at servocity.com/bogie.

2. ADD THE BODY BASE
Attach one 3¾" piece of channel to the center set of mounting holes on the rover, using a large square screw plate and four screws added from the underside (Figure **E**).

3. ATTACH THE HEAD SERVO
Use the standard servo plate to attach the servo motor to the channel. While you're at it, you can attach the coupler that attaches the servo's shaft to the D-rod. Figure **F** shows how it should look.

4. ATTACH THE BODY CHANNEL
Secure the 9" channel to the 3¾" piece you already attached. Use a dual screw plate as shown in Figure **G**.

5. ADD THE SUPPORT BEAM
Attach the second 3¾" channel (Figure **H**) at the top of the 9" piece using the second dual screw plate.

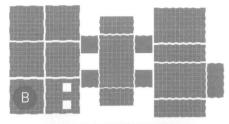

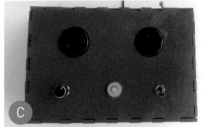
B

C

D

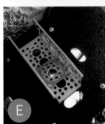
E

F

G

H

I

TIME REQUIRED:
A Weekend

DIFFICULTY:
Moderate

COST:
$200–$300

MATERIALS
» **Plywood: ¼" (8"×8") and ⅛" (about 16 sq. ft)** or cardboard for an easier build
» **Actobotics Bogie Runt Rover Kit** ServoCity #637162, servocity.com. Includes indestructible ABS top plate, six knobby off-road wheels, and six motors to turn them.
» **Standard servo plate B** ServoCity #575124
» **Servo shaft coupler, ¼"** ServoCity #HSA250
» **Bearing, quad pillow block, ¼" bore** ServoCity #535130
» **Setscrew hub, ¼" bore** ServoCity #545548
» **D-shaft, ¼" diameter, 12" long** ServoCity #634094
» **Actobotics aluminum channels, 1½":** 3¾" long (2) and 9" long (1) ServoCity #585443 and 585450
» **Screw plates: large square (1), small square (1), and dual (2)** ServoCity #585430, 585478, and 585472
» **Socket-head screws, #6-32** various lengths
» **Arduino Uno microcontroller boards (2)** One for the robot, one for the controller.
» **Adafruit Motor Shield** Adafruit #1438, adafruit.com
» **NeoPixel Jewel RGB LED boards (2)** Adafruit #2226
» **Servo motor, high torque** A no-name servo such as Adafruit #1142 does the trick.
» **Servo extension cable** Adafruit #973
» **Hookup wire** I'm a fan of SparkFun's wire assortment, #11367, sparkfun.com.
» **Arcade buttons (2)** I used standard buttons, like Adafruit #473 except mine are black.
» **Switches, SPST (2)** standard single-pole, single-throw toggles, like SparkFun #9276
» **Potentiometer** SparkFun #9939
» **Mini breadboard-style PCB** Adafruit #1214
» **RGB LED** SparkFun #105
» **Resistors: 10kΩ (3) and 220Ω (2)**
» **Three-strand wire, 6' or longer** to connect controller to robot. You can use servo wire (ServoCity #57417), or cut a 10-wire ribbon cable (SparkFun #10647) down to three.
» **Paint, green and black**
» **Batteries: 9V (2), AA (4)**
» **9V battery holders, with 5.5mm/2.1mm barrel plug (2)** such as Adafruit #67 or 80
» **Battery pack, 4xAA**
» **Machine screws and nuts, #4**
» **Wood screws, #4**

TOOLS
» **Computer and Arduino IDE software** free download at arduino.cc/downloads
» **Laser cutter (optional)** or saw the shapes out of wood, cardboard, or other materials
» **Soldering iron and solder**
» **Drill and bits**
» **Screwdriver**
» **Wood glue**
» **Paintbrushes**

J

K

L

M

N

O

P

John Baichtal

6. ATTACH THE BEARING

Attach the bearing so that it lines up with the servo's shaft. Figure I on the previous page shows how it should look.

7. SECURE THE D-ROD

Thread the D-rod through the bearing and secure it in the servo's coupler, as shown in Figure J.

8. ASSEMBLE THE FEET AND BODY SKINS

The Creeper's skin consists of a series of laser-cut wooden box shapes that rely on gravity to stay on the robot, as seen in Figure K. The skins are all made from ⅛" plywood, except the head base is ¼".

You can find the blueprints among the downloads for my book at github.com/n1/MinecraftMakers, in the Chapter 9 folder.

9. PAINT IT!

It's time to paint the body and feet that delightful Creeper green. Figure L shows my creation with one coat of paint.

10. ATTACH THE SKIN

Once the paint is dry, drop the skin down on the robot (Figure M) with the ¼" rod projecting from the top. It should fit nicely without any problem.

11. ADD THE HEAD BASE

Use ¼" plywood, 8"×8" square. Drill a ¼" center hole as well as the mounting holes for the setscrew hub; then secure the hub with the small square screw plate. Figure N shows the base.

12. BUILD THE HEAD

Assemble the panels of the head (Figure O). Once again, a plywood box! This one needs holes for eyes.

13. PAINT THE HEAD

Paint the outside of the head green to match the skin, but also add the black parts of the face. I suggest painting the inside of the head black (Figure P) so the eyes look blacker.

14. INSTALL THE ARDUINO

Find a spot on the underneath of the robot to install the Arduino using #4 hardware (Figure O). You can drill into the Bogie's ABS chassis if you want, or use one of the available mounting holes.

15. SEAT THE MOTOR SHIELD
The motor shield (Figure R) sits right on top of the Arduino, with the shield's male headers plugged into the Arduino's female headers.

16. CONNECT THE SERVO
Attach the servo wires to the motor shield's Servo 1 pins, as shown in Figure S.

17. MOUNT THE 9V BATTERY
Attach the 9V battery to the chassis but don't plug it in yet. (Figure T shows it already plugged in.) This powers the Arduino but not the motors.

18. ATTACH THE BATTERY PACK
Attach the 4xAA battery pack to the chassis and plug it into the motor shield's Power terminals, as shown in Figure U. This pack powers the motors separately.

19. CONNECT THE MOTORS
You have six motors, three on the left and three on the right. Combine the leads as you see in Figure V, so that the M3 motor terminals control one side, and M4 controls the other. The Bogie's motors are modest in size and stacking them won't strain the motor shield's capabilities.

20. WIRE UP THE NEOPIXEL EYES
You're using two NeoPixel Jewels for eyes, to create the telltale glowing red that signals an imminent explosion. Connect both Jewels' VIN (red wire) and GND (black) pins to the Arduino's GND and 5V pins. The data wire goes from digital pin 6 on the Arduino to the IN pin on the first Jewel, and then from OUT to IN on the next eye (Figure W).

21. WIRE THE CONTROLLER CONNECTIONS
You'll need three wires to attach the Creeper to its controller. As you can see in Figure X, one wire connects to digital pin 0 on the Arduino, another to pin 1, and a third to GND. You'll connect these to their counterparts on the controller's Arduino. You can make these wires as long as you want, but 6' is probably good enough.

PROGRAM THE CREEPER
The Creeper is a simple robot, and this is reflected in the code. Turn the page for the complete code listing:

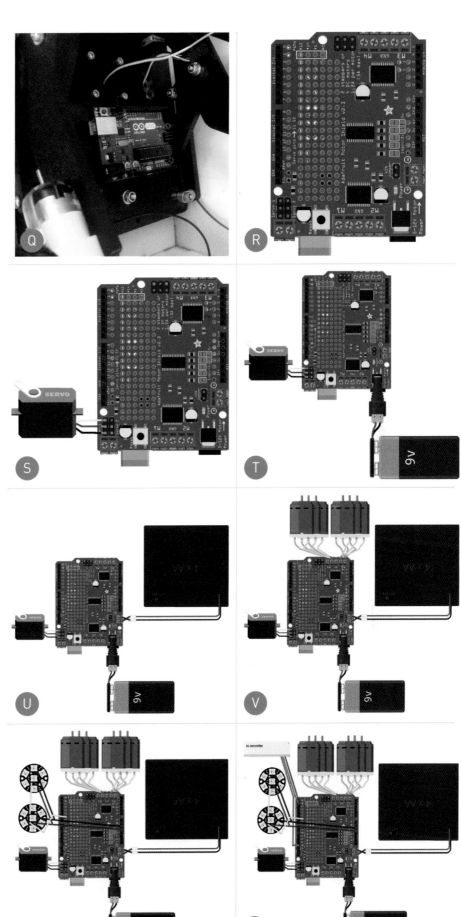

```
#include <Wire.h>
#include <Adafruit_MotorShield.h>
#include "utility/Adafruit_PWMServoDriver.h"
#include <Servo.h>
#include <Adafruit_NeoPixel.h>

Adafruit_MotorShield AFMS = Adafruit_MotorShield();
Adafruit_DCMotor *leftMotors = AFMS.getMotor(3);
Adafruit_DCMotor *rightMotors = AFMS.getMotor(4);
Servo servo1;

#define PIN 6
Adafruit_NeoPixel strip = Adafruit_NeoPixel(14, PIN,
NEO_GRB + NEO_KHZ800);

const int buzzerPin = 13;

void setup()
{
    Serial.begin(9600);
    AFMS.begin();
    servo1.attach(10);
    strip.begin();
    strip.show(); // Initialize all pixels to 'off'
}

void loop() {
    if (Serial.available() >= 2)
    {
        char start = Serial.read();
        if (start != '*')
        {
            return;
        }
        char cmd = Serial.read();
        process_incoming_command(cmd);
    }
    delay(50); //limit how fast we update
}

void process_incoming_command(char cmd)
{
    int speed = 0;
    switch (cmd)
    {
    case 0:
        //DO NOTHING
        break;
    case 1:
        //DRIVE LEFT WHEELS
        leftMotors->setSpeed(200);
        leftMotors->run(RELEASE);
        break;
    case 2:
        //DRIVE RIGHT WHEELS
        rightMotors->setSpeed(200);
        rightMotors->run(RELEASE);
        break;
    case 3:
        //TURN ON EYES
        strip.setPixelColor(0, 255, 0, 0);
        strip.setPixelColor(1, 255, 0, 0);
        strip.setPixelColor(2, 255, 0, 0);
        strip.setPixelColor(3, 255, 0, 0);
        strip.setPixelColor(4, 255, 0, 0);
        strip.setPixelColor(5, 255, 0, 0);
        strip.setPixelColor(6, 255, 0, 0);
        strip.setPixelColor(7, 255, 0, 0);
        strip.setPixelColor(8, 255, 0, 0);
        strip.setPixelColor(9, 255, 0, 0);
        strip.setPixelColor(10, 255, 0, 0);
        strip.setPixelColor(11, 255, 0, 0);
        strip.setPixelColor(12, 255, 0, 0);
        strip.setPixelColor(13, 255, 0, 0);
        strip.show();
        break;
    case 4:
        //TURN OFF EYES
        strip.setPixelColor(0, 0, 0, 0);
        strip.setPixelColor(1, 0, 0, 0);
        strip.setPixelColor(2, 0, 0, 0);
        strip.setPixelColor(3, 0, 0, 0);
        strip.setPixelColor(4, 0, 0, 0);
        strip.setPixelColor(5, 0, 0, 0);
        strip.setPixelColor(6, 0, 0, 0);
        strip.setPixelColor(7, 0, 0, 0);
        strip.setPixelColor(8, 0, 0, 0);
        strip.setPixelColor(9, 0, 0, 0);
        strip.setPixelColor(10, 0, 0, 0);
        strip.setPixelColor(11, 0, 0, 0);
        strip.setPixelColor(12, 0, 0, 0);
        strip.setPixelColor(13, 0, 0, 0);
        strip.show();
        break;
    case 5:
        //TURN HEAD 90DEG LEFT
        servo1.write(0);
        delay(15);
        break;
    case 6:
        //TURN HEAD 45DEG LEFT
        servo1.write(45);
        delay(15);
        break;
    case 7:
        //TURN HEAD FORWARD
        servo1.write(90);
        delay(15);
        break;
    case 8:
        //TURN HEAD 45DEG RIGHT
        servo1.write(135);
        delay(15);
        break;
    case 9:
        //TURN HEAD 90DEG RIGHT
        servo1.write(180);
        delay(15);
        break;
    }
}
```

You can download this code file, *creeper.ino*, at github.com/n1/MinecraftMakers in the Chapter 9 folder. Then open it in the Arduino IDE, and Upload it to the Arduino.

The sketch works by listening for commands from the controller via the serial connection, then activating the appropriate LEDs or motors. Of course you can modify the code to fine-tune your mob or give it new behaviors.

BUILD THE CONTROLLER

The controller I built is a simple plywood box 6"×4"×2", and it fits nicely in my hand.

1. ASSEMBLE THE BOX

As usual, I'm including the vectors for my controller in with the downloads. I painted the box purple just for fun.

2. INSTALL THE ARDUINO

Use wood screws to attach the Arduino to the bottom of the controller box (Figure Y).

3. WIRE UP THE BUTTONS

Take the mini breadboard PCB and attach the two buttons and the switch that turns on the eyes. They all work the same way in Arduinoland. Connect one lead to power (shown as green wires in Figure Z). Connect the other leads (gray wires) to the Arduino digital pins 9 (switch) and 5 and 6 (buttons), and also to GND via a 10K resistor.

4. ATTACH THE POTENTIOMETER

The first lead connects to power, the third lead connects to GND, and the middle one to analog pin A0 on the Arduino. Figure AA shows the pot wired up.

5. ADD BATTERY AND POWER SWITCH

The second switch controls power to the Arduino. Wire it in line with the 9V battery (Figure BB) — or skip it if your battery pack already has a switch.

6. INSTALL THE LED

Grab your RGB (tricolor) LED. Ignore the blue lead. Connect the red and green leads to digital pins 10 and 11, respectively, with 220-ohm resistors protecting the LED, and the common cathode to GND (Figure CC).

7. CONNECT TO THE ROBOT

Connect the three long wires from the robot to the same pins on the controller (Figure DD).

8. PROGRAM THE CONTROLLER

Download *purple_controller.ino* from the GitHub page and upload it to the Arduino.

GET CREEPIN'

Your mob is ready to go mobile. Press the arcade buttons to steer the robot base, while turning the head with the potentiometer. Now activate the robot's menacing red eyes by flicking the switch! ●

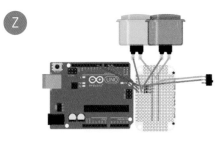

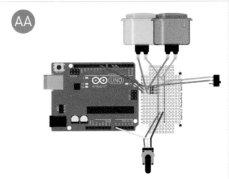

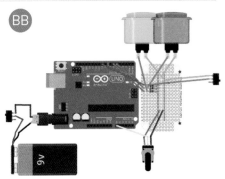

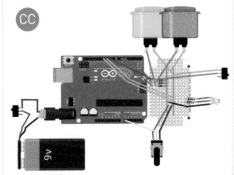

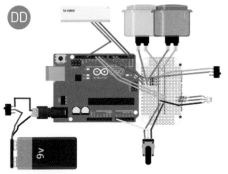

John Baichtal, Mojang

Go, Dog. Go!

Build a custom wheelchair for a pet in need

Written by Erica Charbonneau

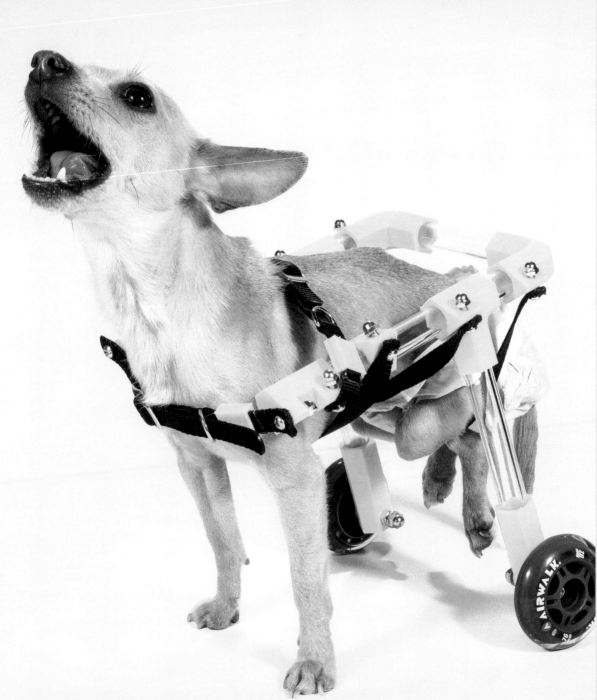

TIME REQUIRED:
A Weekend

DIFFICULTY:
Easy/Moderate
(but beginner friendly, provided you have access to a 3D printer)

COST:
$50–$75

MATERIALS

- » **3D printed pieces (10)** from thingiverse.com/thing:1397964, printed in PLA or ABS
- » **Acrylic or aluminum tubes (5–6)** 5 tubes for a dog 10 pounds or less, otherwise 6
- » **Wheels (2)** such as Razor scooter wheels, Amazon #B000FDFCPA amazon.com. Larger wheels are recommended for dogs over 15 pounds.
- » **Skate bearings, standard (2)** if not included with the wheels
- » **Dog collars (2)** slightly larger than dog's size
- » **Dog leashes or webbing** I recommend two leashes for a large dog.
- » **Small screws, 10-24 × ¾", or preferred size (10)**
- » **Large screws, 10-24 × 3" (2)** for securing wheels on piece E
- » **Acorn nuts, 10-24 (12)**
- » **Fleece** or preferred padding material

TOOLS

- » **3D printer (optional)** To find access to a machine or a printing service you can use, check out makezine.com/where-to-get-digital-fabrication-tool-access.
- » **Screwdriver**
- » **Adjustable wrench**
- » **Scissors**
- » **Measuring tape**
- » **Lighter**
- » **Soldering iron (optional)**
- » **Gorilla Glue (optional)**

A friend needed a wheelchair for their French Bulldog at short notice. They were unable to afford the costly commercial wheelchairs available online. I rose to the occasion to design something for an adorable dog, and the results were fantastic. Murray loved her chair!

I've since built two other chairs for dogs, and I've worked on iterating on what I call the "FiGO" design and documentation to encourage dog owners to tackle this project for their pet in need.

This device uses parametrically designed 3D printed joint pieces that fit into acrylic or aluminum tubing. The tubing can be easily customized to the dog for both fit and style, and the 3D printed pieces can also be personalized via the Customizer application on Thingiverse. Currently screw size, tube outer diameter, wheel angle, and your dog's measurements can be input to view a rendering of your dog's custom wheelchair. However, the standard pieces are available for download on Thingiverse at thingiverse.com/thing:1397964. I'm working on making the pieces resizable based on your dog's weight for added strength (currently you can do this by manually scaling the pieces).

All the other materials used in the project can be sourced locally at a hardware store or online on amazon.com.

1. MEASURE YOUR PET

In order to determine what size your wheelchair frame needs to be, you'll have to measure your pet properly. The diagram in Figure ❶ depicts the three measurements that need to be taken. Measurement **A** is the width of your pet, measurement **B** is the height of your pet from the ground to the top of their shoulder bone, and measurement **C** is the length of your pet from chest to tail.

Once you have these measurements, you'll need to do a few simple calculations to determine the lengths of tubing that will be required for your FiGO wheelchair. You will need 6 pieces of tubing, cut to the following lengths:

ERICA CHARBONNEAU is a creative technologist and inclusive developer. She currently holds a Master of Design, Inclusive Design from OCAD University in Toronto, Ontario. She is interested in how communities design, develop, fabricate, and share downloadable assistive technologies online.

IMPORTANT: Consult with your vet if you have any questions or concerns about the fabrication or fit of the wheelchair.

» 2 width tubes: measurement **A** plus 1"
» 2 length tubes: measurement **B** minus the radius of your chosen wheel, then minus 1"
» 2 height tubes: measurement **C** minus 2"

I have created a simple Google spreadsheet (makezine.com/go/pet-wheelchair-calculator) that does these calculations for you as you input your dog's measurements. Click on File→Save a Copy to edit the file for yourself.

2. 3D PRINT THE JOINTS

Print all 10 of the joint pieces, **A**–**E** (Figure ❷). You'll print 2 each of **A**, **C**, **D**, and **E**; these pieces are identical on either side of the wheelchair.

The **B** piece comes in two versions, mirrored for left and right sides. Print one of each (both files are provided on Thingiverse in the files section).

3. BUILD THE FIGO FRAME

Take both of your **A** pieces and a width tube, and slide an **A** piece onto each end of the tube. Make sure the tabs that will be used for the straps are facing in the direction pictured in Figure ❸.

Next connect both **A** pieces to both length tubes (Figure ❹).

Slide both **B** pieces down the length tubes leaving about 1" or so between **B** and **A** (this will change depending on your chair size, and you can easily adjust these pieces later to match your pet's size). Make sure the tabs on both pieces are facing outward, and are both pointing to the back of the wheelchair (toward piece **A**) as shown in Figure ❺.

Now do the same with both **C** pieces, remembering that the tabs are on the outside of the wheelchair (Figure ❻).

Cap both **D** pieces on the ends of the length tubes, with the tabs facing outward (Figure ❼).

Place both height tubes (measurement **B**) in both **B** pieces (Figure ❽) and then set your wheelchair frame aside.

Now take both **E** pieces and the remaining width tube and connect them together (Figure ❾, following page). (For dogs

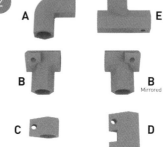

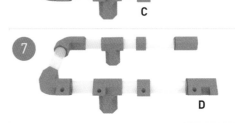

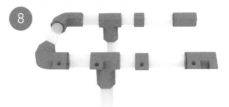

smaller than 10 pounds, this step is not required; you can print a different version of piece **E** with no added support bar.)

Connect both **E** pieces to the bottom of the height tubes on your frame (Figure 10).

Make sure you've added your bearings to your wheels if they don't already have them. Then bolt the wheels to the **E** pieces using the 3" screws, capped with acorn nuts (Figure 11). You may need to use a screwdriver and/or a wrench for this.

Finally, add the straps (Figure 12). Both collars will be fastened to pieces **D** at the front of the chair, and the rest of the straps will be made out of a dog leash (or any piece of webbing that you may have). Pieces **A** will require a longer strap, for the dog's legs. Pieces **B** and **C**'s straps are for the dog's belly and will need to be the same size.

Cut the dog collars in half (Figure 13); you'll use the existing buckles to adjust the fit. Later you can cut them down shorter if need be. The fit will have to be very secure.

To measure the body straps, first place the leash around the dog when it's standing in the chair, to see how long these need to be. Make the leg strap slightly longer than the belly straps.

Cut the straps with scissors, then quickly pass over the ends with a lighter to prevent fraying. Once all your straps are cut to size, punch a hole into each end with scissors (the collar straps only need to be punched at the ends that you cut). Better yet, use a soldering iron to burn a hole through the strap (Figure 14). This will melt the plastic, preventing any fraying.

Secure the straps to the tabs on each joint piece (Figure 15) using the ¾" screws and acorn nuts, and make sure they're tight. (I used a different type of screw in the prototype shown here.)

USING THE CHAIR

All dogs will take time to get used to their wheelchair. Some adapt really quickly and some hate their chairs initially. It takes work to train a dog to be comfortable with their wheelchair.

I've worked with three dogs so far and I've experienced: a very comfortable and quickly adapted dog, an anxious and moody dog, and a timid pup. I've learned that leaving the chair in your living space and just letting your dog approach it and sniff it helps! ✏

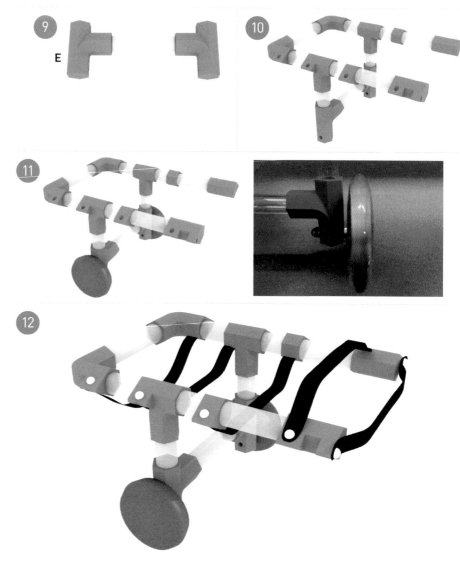

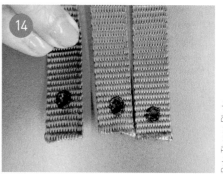

Pete Thorne Photo

NOTE: After the FiGO wheelchair is complete, the width tube between the two **A** pieces, as well as the height tubes connecting pieces **B** to **E**, can be glued with Gorilla Glue to ensure stability of the chair. This is not as important when the dog is very light, but is recommended otherwise.

Toy Inventor's Notebook

VARI-TONE MINI-CAJON Invented and illustrated by Bob Knetzger

Already made a cigar-box ukulele? Here's a sister project that adds percussion to the band: a mini cajon drum. It has a special musical feature I call the "Vari-Tone." When you squeeze the cajon between your knees, the bottom opens and closes to change the sound: closed for a lower/dark tone, or open for a brighter/thin sound, or anything in between. It's like a wah-wah pedal for your cigar box cajon!

A spring-loaded paddle is connected to a simple linkage joined to the box's lid. As the paddle is pressed inward, the bellcrank swivels to push the lid open (Figure Ⓐ). There's another swivel joint parallel to the lid hinge that keeps the motions smooth. A spring swings the lid closed again.

Figure Ⓑ is an exploded view of the mechanism. I shaped a wood dowel with a saw and file to make the links, and cut a piece of 0.090" styrene plastic for the bellcrank. I threaded the bellcrank's holes for easy assembly with small screws but you could also use washers and nuts. The other swivel joints are pinned with short bits of press-fit drill rod, but small wood screws would work too. Raid your parts bin for a coil spring with the right amount of springiness and big enough to go around the dowel. Drill a hole in the dowel to pin the spring to it with a bit of wire, so that the spring pushes against the outside of the box when assembled.

Choose a cigar box that has a real wood bottom for a good tone. I vacuum-formed a small paddle with a curved shape to fit my leg and taped it to the side of the box as a hinged paddle to depress the spring. ✱

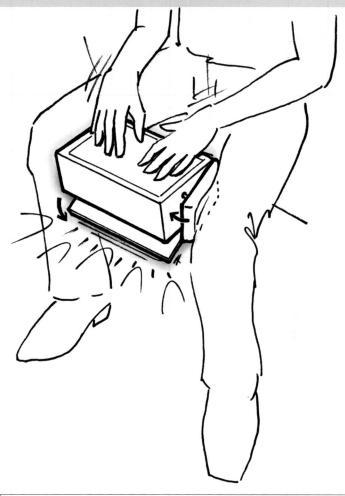

MATERIALS
- » **Cigar box with wooden bottom**
- » **Wood dowel, ⅜" diameter**
- » **Plastic, about ⅛" thick**
- » **Small machine screws, nuts, and washers** If you thread the bellcrank you can delete the nuts and washers.
- » **Small wood screws**
- » **Coil spring, compression type** to fit over the dowel
- » **Plastic or wood** for making the paddle
- » **Bit of wire**

TOOLS
- » **Drill and bits**
- » **Handsaw or band saw**
- » **File**
- » **Utility knife**
- » **Screwdriver**
- » **Tape**
- » **Thread tap (optional)**

BOB KNETZGER is a designer/inventor/musician whose award-winning toys have been featured on *The Tonight Show*, *Nightline*, and *Good Morning America*. He is the author of *Make: Fun!*, available at makershed.com and fine bookstores everywhere.

Ⓐ

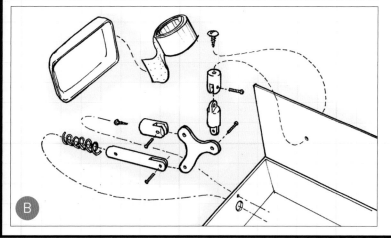

Ⓑ

Corgi Keyboard

Written by Ashley Qian

How I made my first musical plushie

ASHLEY QIAN
is a software
engineer by day,
artist by night,
kid at heart.
ashleyqian.com

Corgi Keyboard V2 on the left and
Giraffe Guitar V1 on the right.

DIY.org has a tradition during the holidays called "DIY Secret Santa," where we create a present for an assigned co-worker. It's one of my favorite traditions because it's a great opportunity to learn new skills. In 2016, I was Adam's Secret Santa (Figure A).

That year, I had been itching to learn electronics, so I decided to make Adam a Corgi-shaped electric keyboard.

REFINING THE IDEA

I had so many questions. Where would the sound come from and how would I trigger it? If it was going to be an electronic instrument, what kind of logic board should I use? What material would the Corgi be made out of? How would the Corgi enclosure integrate with the electronics?

I was getting overwhelmed by all the knowledge that I didn't have, so I tried to break the problem down into smaller components. I started with the most obvious observation: The Corgi had to play some sort of sound when a part of its body was pressed (Figure B).

Once I knew all of the problems that I needed to design for, I began to answer each unknown one by one.

How does a set of hands "play" a Corgi?
A simple button (Figure C) seemed like the easiest way to trigger a sound, but I thought they would look awkward strewn across the body. If only I could figure out a way to cover it up …

Since I was trying to make a Corgi keyboard, it made a lot of sense to follow the visual aesthetic of a piano. I could create a mechanism where the button would be triggered by pressing on something that seemed like a piano key (Figure D). This meant that the Corgi enclosure would have to be made out of something hard: like wood, cardboard, or acrylic.

As much it'd be fun figuring out how to use a laser cutter to cut acrylic pieces or starting a woodworking learning adventure, I determined that I needed an alternative design that employed materials that I was more familiar with. In that moment, I realized that fingers were not only good for poking and pressing things — they are also fairly conductive (Figure E)!

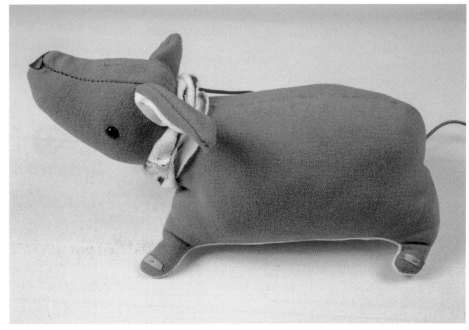

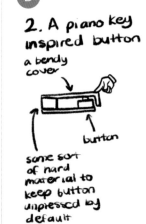

A — Meet Adam!
he is a software engineer
he likes Corgis
he likes music production

B — Design Problem #1
Corgi must have a touchable trigger

Design Problem #2
Corgi must translate the poke to which sound it should play

Design Problem #3
Corgi must play the sound

C — 1. A button

D — 2. A piano key inspired button
a bendy cover
button
some sort of hard material to keep button unpressed by default

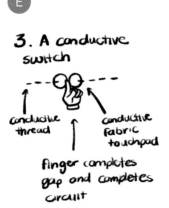

E — 3. A conductive switch
conductive thread
conductive fabric touchpad
finger completes gap and completes circuit

With this design, I could make a squishy Corgi plushie with touchpads that would trigger musical notes.

How does the Corgi interpret, then translate that input into sound?
The conductive touch triggers help the Corgi determine whether or not it should play a sound. With the power of a computing board, the Corgi can also determine which sound to play. To figure out which board to use, I did a bit of research and listed the attributes of a couple of boards that I knew (Figure F).

I decided to go with the Makey Makey since it was the quickest way to get to a working prototype. All I had to do was clip the alligator clips to a conductive surface, plug the Makey Makey into a computer, and navigate to a website that played sounds when different triggers go off (Figure G).

SEWING AND DESIGNING A PLUSHIE

I had never sewed or designed a plushie before, so I tried to find as many dog plush patterns as possible. The best places to research these patterns were on Google Images and Pinterest. After prototyping the dog patterns with pieces of scrap paper and tape, I started to notice a trend to all of the patterns that I found. I took what I learned and made adjustments on top of the basic concepts to create a pattern of my own (Figure H).

G
Diagram inspired by the Makey Makey tutorial.

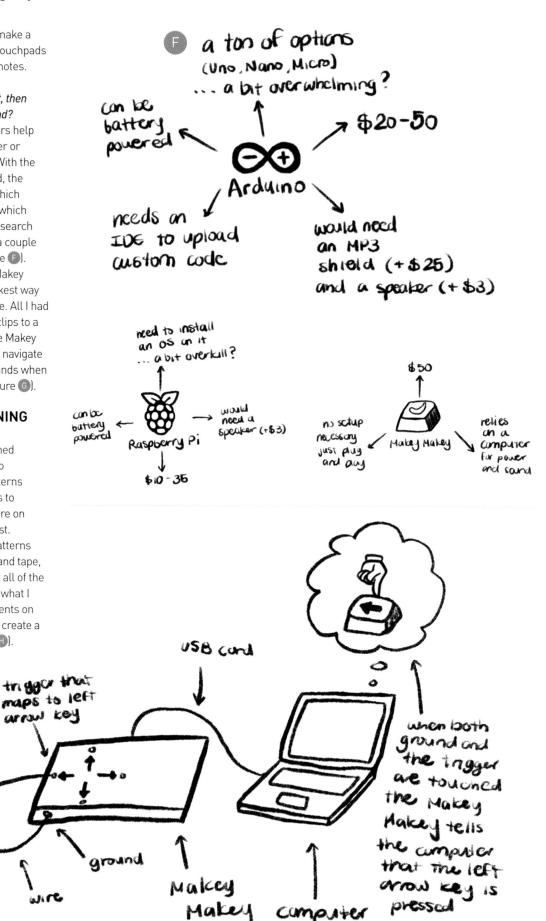

H

Observation #1

Most 4 legged animal plush designs are triangular prisms... with legs and a head

triangular prism | add feet | add a cute corgi face

Observation #2

Most dog plush patterns have what's called a crown, to give depth to the head. The wider you make the crown the fuller the head will be

no crown | with crown

My Pattern

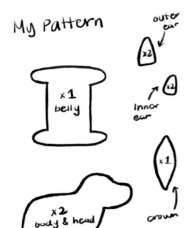

outer ear ×2

inner ear ×2

×1 belly

×1 crown

×2 body & head

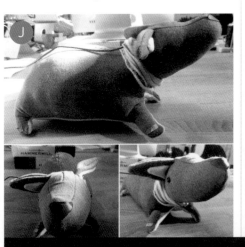

I

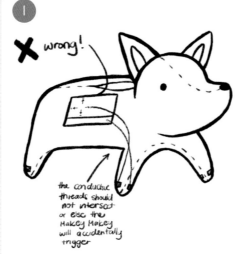

✗ wrong!

the conductive threads should not intersect or else the Makey Makey will accidentally trigger

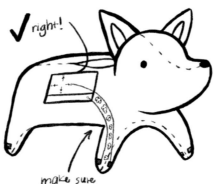

✓ right!

make sure to put stuffing in between the threads to prevent them from touching

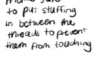

THE INTERSECTION OF ELECTRONICS AND FABRIC

The final step was to figure out how to connect the Makey Makey to my conductive-tape touchpads on the two sides of each paw. I knew that the Makey Makey used alligator clips to connect the conductive item to the board, so I assumed that conductive thread wouldn't be so different.

Boy, was I wrong. I quickly realized my mistake when the first string on one side (ground) of the paw touched the second string on the other (trigger). The alligator clip wires were insulated with plastic, and my conductive thread was not. This meant that every time two threads touched, the Makey Makey would trigger a keypress, even when my fingers weren't pressing the touch points! After hours of carefully stuffing cotton in between the threads, I managed to get them in a configuration where they weren't touching each other (Figure Ⓘ).

HINDSIGHT IS 20/20

It's been a year since I made my first musical plushie (Figure Ⓙ). Since then, I've made him a couple of friends to play with (Figure Ⓚ), and together they debuted at two Maker Faires (Bay Area and New York).

Looking back, I made many changes from the original design. I ended up swapping the Makey Makey for an Adafruit Feather BlueFruit LE for cost and customization reasons. I added a zipper for accessing the board, and I wrapped the conductive thread with heat-shrink to prevent false triggers. Now I'm using wires soldered to snaps, so the electronics module can snap in and out of the plushie.

I've learned so much from this journey. It's crazy to remember that a year ago, I knew nothing about electronics or plushie design. But, I wouldn't have gotten to where I am today if I didn't allow myself to try, experiment, or make mistakes.

I can't wait to find out what more I can create. ●

See more on Maker Share: makershare.com/projects/corgi-keyboard-v1

Dragonfly Helicopter

Scratch-build a wind-up flyer that darts into the sky!

Written by Slater Harrison and Rick Schertle

Hep Svadja, Rick Schertle

TIME REQUIRED:
30–60 Minutes

DIFFICULTY:
Easy

COST:
$2–$3

MATERIALS
- » Tape, clear
- » Drinking straws
- » Cotton swabs with plastic sticks
- » Plastic bottle, 2-liter
- » Paper clip
- » Plate, foam
- » Cardboard, scrap
- » ⅛" super sport rubber from faimodelsupply.com. You'll need 11" per copter.
- » Printed 8½"×11" template from makezine.com/go/dragonfly-helicopter-pattern. Print it full-scale, not fit-to-page.

TOOLS
- » Ruler
- » Scissors
- » Pushpin
- » Paper punch, 1-hole
- » Glue gun, 10-watt Lower temperature is best.
- » Long-nose pliers
- » Razor blade

Q. Why did you make the helicopter look like a dragonfly?
A. I didn't — at least not intentionally. I set about to make the most efficient, highest-flying helicopter. The wings serve a critical purpose: to keep the fuselage from spinning too much, so the propeller can spin more. I discovered that long, thin wings attached near the top worked best! —*Slater Harrison*

RICK SCHERTLE
runs the Maker Lab at Steindorf K-8 STEAM School in San Jose, California. He's the author of: *mBot for Makers* and co-founder of AirRocketWorks.com.

SLATER HARRISON
learned creative engineering by making machines for village industries in Bangladesh. He learned patience as a middle and high school technology teacher for 28 years.

ROTOR HEADS

I met Slater Harrison (aka Science Toy Maker, sciencetoymaker.org) a few years ago online. Like-minded in a number of ways, we connected via email and have kept up ever since. Slater is a teacher, like me, and his brilliant science projects use off-the-shelf materials that are affordable for budget-minded public school teachers. I was especially drawn to his flying projects, and featured his stomp rocket design in my book *Planes, Gliders, and Paper Rockets* (published by *Make:*). Some of my favorite flying projects are low-tech, work well with large groups, and are inexpensive. The Dragonfly Helicopter fits the bill!

I first came across the Dragonfly Helicopter at World Maker Faire New York a couple years ago. Recently, Slater has perfected the design of the reliable rubber band helicopter featured here. "There is something empowering about making a propeller from a trashed 2-liter drink bottle that's superior to commercial propellers," he says. "Your flying machine will go higher than the tallest trees with no batteries. Unlike model airplanes, which can be plagued by stability problems from a slightly warped or crooked wing, helicopters always fly true. Nobody is excluded. Windup helicopters are so inexpensive that whole classes of students can make them for much less than a dollar each."

I built these with my middle schoolers and they loved how the copters whimsically shot up into the air, and how they were able to easily tweak the design to see how the flights changed. With a bit of time and some simple supplies, you'll be having high-flying fun too!
—*Rick Schertle*

1. PRINT THE TEMPLATE

Find the template at makezine.com/go/dragonfly-helicopter-pattern and print it on a standard 8½"×11" piece of paper. Use a ruler to make sure it's printed full-scale, not "fit-to-page" from your printer menu. For this fixed-pitch model of the Dragonfly, you'll use all of the features on the template.

2. BUILD THE PROPELLER HUB ASSEMBLY

First, bend the propeller shaft. Straighten a small paper clip, mark the wire with a Sharpie at 5cm, and then cut it off with the inner, wire-cutting part of your long-nose pliers. Using the tip of the pliers, bend a hook into the diamond shape shown on the template (Figure Ⓐ). To get tight, sharp bends, apply pressure on the wire right next to the pliers (Figure Ⓑ).

Next, cut the propeller hub. Cut a 5cm length of plastic tube from a cotton swab stick. Mark the center (2.5cm), then poke a hole through the center using a pushpin (Figure Ⓒ). For safety, stick the pin into a scrap of cardboard and set it aside.

Make a bearing for your propeller shaft by cutting 13mm from another swab stick. Then make a washer by hole-punching a piece of 2-liter bottle plastic then piercing a hole in the center of it with a pin (Figure Ⓓ).

Slip the pieces onto the wire shaft in this order: bearing, then washer,

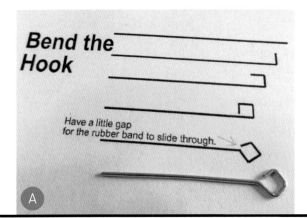

Bend the Hook

Have a little gap for the rubber band to slide through.

Ⓐ

Ⓑ

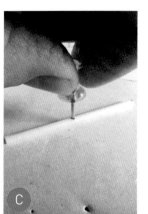

Ⓒ

Ⓓ

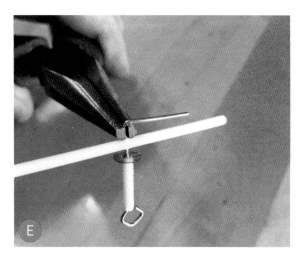

E

Fixed pitch gauge

bottom

F

G

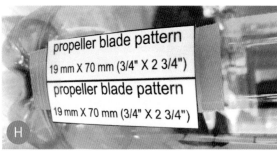

propeller blade pattern
19 mm X 70 mm (3/4" X 2 3/4")
propeller blade pattern
19 mm X 70 mm (3/4" X 2 3/4")

H

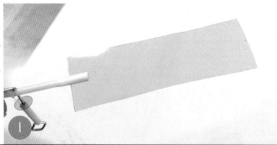

I

then hub stick. Then bend the end of the wire 90°, leaving wiggle room for the shaft to turn in the plastic bearing (Figure E). Cut the bent wire off at about 1cm and tape it to the hub stick with clear tape.

3. CREATE PITCH AND ADD BLADES

Now we're going to fix the pitch (angle) of our helicopter rotor by crushing the hub stick and cutting slits to hold the plastic blades. The pitched blades allow the propeller to "grab" the air and pull the copter up.

Cut out the fixed pitch gauge from the template and tape it to bottom jaw of your long-nose pliers (Figure F). Pinch the hub stick at the 22½° angle of the pitch gauge, and squeeze it tight. Do this on both ends.

Now carefully cut a slit right on the folds at each end of the smashed swab stick, so you end up with two flaps (Figure G). Wear a glove or use a rag to protect your fingers holding the propeller assembly.

Cut out the propeller blade pattern from the template, tape it to the smooth section of a 2-liter bottle (Figure H), and then cut out the plastic to make two blades. (For doing many blades with a group, use the template at makezine.com/go/many-blades-pattern.)

Slip the blades into the slits on your propeller hub, with the concave side of the blades facing down (Figure I). When you're sure they're in line with the propeller hub, tape the blades in place on both sides. Trim the tape, and also clip a tiny bit off the sharp corners of the blades for safety. Now hold the hub, and your propeller should turn like a windmill when moved through the air!

4. TIE THE RUBBER LOOP

Cut off 11" of rubber band. Put both ends together and tie in an overhand knot (see makezine.com/go/rubber-knot for a detailed illustration). Tighten it a little, then pull on the rubber band loop to move the knot down to the very bottom of the loop (Figure J). Set it aside for later.

IMPORTANT: Office rubber bands don't hold much energy at all. For your Dragonfly to go high, you need "sport" or "hobby" rubber bands specially made for rubber band airplanes.

5. MAKE FUSELAGE AND ATTACH PROP

Cut off two drinking straws at 5½" as shown on the template. (If you're using flexible straws, this is about the length of the long section before the flex.) Tape the straws together at the top and bottom with clear tape.

Cut off two 1cm × 0.5cm scraps of foam plate. Sandwich them together with glue, then glue them to the end of the straws as shown in Figure K. This creates a spacer between the propeller assembly and fuselage.

Now grab the propeller hub assembly and glue the bearing to the spacer as shown. It's very important that you don't get glue in the propeller, and that the propeller blades and the shaft hook spin freely.

6. ADD WINGS AND RUBBER BAND

Cut a rectangular piece of foam plate 10cm × 4.5cm for the wings. Slide the wings between the two straws about 25mm (1") from the top of the fuselage, center it, and tape it in place.

Without wings, the fuselage just spins too fast and it doesn't fly. Bending the wings upward gradually is called *dihedral* and it adds flight stability. Splitting the wings for the four-winged "dragonfly" look is optional. You can also decorate them.

Cut the straws at the bottom of the fuselage at a slight angle for the rubber band holder, as shown in Figure L. Then, cut a scrap of swab stick to 2.5cm and glue it to the bottom of the fuselage.

Finally, hook the rubber band onto the propeller hook and then around the swab stick holder on the bottom of the fuselage so the knot is at the bottom as shown.

Your helicopter is complete! ✏

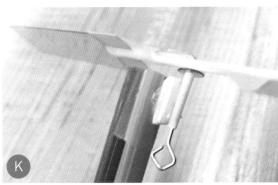

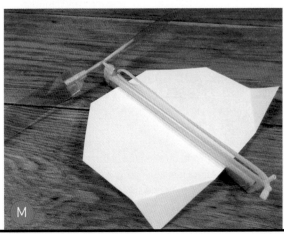

NOW GET YOUR ROTOR RUNNIN'

Fly and Adjust!

To get your copter airborne, simply wind the propeller clockwise — up to 150 turns — and let it go!

Your Dragonfly may bounce around a bit in the air because the prop needs to be balanced. To do this, take the rubber band off and turn the Dragonfly sideways so the propeller can hang loosely. The heavier blade will rotate downward. Cut a little off the heavier end or add some tape to the lighter end until the blades are balanced and hold position at any angle.

To make your Dragonfly fly even higher, rub baby shampoo or liquid soap on the rubber band to reduce friction. Don't use petroleum, grease, or oil-based lubricants as these will destroy the rubber.

The Dragonfly Helicopter works well in a room with high ceilings and outside too!

Modding Your Dragonfly

Have fun experimenting with different materials for the wings and fuselage, and different pitch angles for the blades.

Alternative wings — Foam picnic plates can vary greatly in weight, thickness, and density. An alternative to foam wings is paper wings cut from a quarter-sheet of printer paper. Cut off the top corners to avoid flapping and bend the bottom corners for added flight stability (Figure M). It won't have the dragonfly profile, but will fly very stable and consistent.

Foam fuselage — Foam grocery trays can work well for a fuselage also, the kind that meat and produce come on. The angle on the edge can be used as the spacer for the propeller assembly.

Easier build — You can use a craft stick for the fuselage and a store-bought propeller. These copters are heavier and don't go as high, but are easier to build for younger makers.

Variable pitch — Build extra propellers with different pitch angles, or try building the variable-pitch version of the Dragonfly; instructions can be found at the link below. It's slightly more complicated but it lets you vary the pitch of the blades to find the optimum angle for your copter.

More modification ideas and a detailed build video can be found at: sciencetoymaker.org/dragonfly-helicopter.

Curing Cuteness

Embed and embellish electronics in molded resin to craft adorable accessories

Written and photographed by
Rachel "Konichiwakitty" Wong

Mixing resin crafts with electronics is a fun way to explore your creativity. Resin is a strong and hard material once fully cured. Any electronics that have been embedded will be secure and stable from external factors and can be used for all kinds of projects.

Advantages of resin include waterproofing, protection from wear and tear, and cosmetic appearance. Compared to other embedding materials, resin gives a clear shine finish, and just like with 3D printing, the final shape of the resin can be controlled precisely — in this case, by using various silicone molds. This allows us to build anything we want, especially attractive accessories such as watch faces or jewelry like rings and pendants.

The type of resin depends on the silicone mold selected. If an opaque silicone mold is used, polymer resin is best and takes 24 hours to cure at room temperature. UV-cured resins require translucent silicone molds that allow the UV light to penetrate the mold, causing the UV resin to cure completely and almost immediately.

Even after your resin has hardened, the surface may remain tacky for a longer period, therefore it is important to allow more time to dry fully.

1. PREPARE YOUR RESIN

Follow the instructions on the packaging of your purchased resin. It's important to ensure that the correct ratio is mixed so that the resin cures properly.

2. ADD COLORANT AND/OR EMBELLISHMENTS

Once the resin is prepared, mix in any colorant, glitter, or embellishments using the bamboo skewers. For colorants, use only recommended pigments that won't affect the liquid ratio. If your project will be more than one color, use different bamboo skewers for each.

RACHEL WONG (@konichiwakitty) is growing eyes in the lab using stem cells to study and seek cures for blindness as part of her Ph.D. She also tinkers and hacks electronics to create wearable fashion technology .

3. POUR YOUR RESIN INTO THE SILICONE MOLD

Using the bamboo skewer, direct your resin to fill up all the corners of your mold (Figure Ⓐ), ensuring there are no huge air bubbles (big bubbles will cure as a big gap in your project). Use the long reach lighter over the surface of the resin (Figure Ⓑ) to warm it gently and encourage any surface air bubbles to pop.

4. ADD AND ARRANGE ANY SURFACE EMBELLISHMENTS

For the embellishments to appear at the surface of your resin project, use the bamboo skewer to arrange and push the embellishments to the bottom of the mold (Figure Ⓒ). If necessary, use the long reach lighter again to remove bubbles.

5. ADD LEDS

You can add as many LEDs as you like but it is important to remember that the position of the positive and negative legs will affect your final circuit. The fewer the LEDs, the easier it is to complete your circuit. In this step, I have used the bamboo skewers to hold up my LEDs and keep them in position while the resin cures (Figure Ⓓ).

6. SOLDER YOUR CONNECTIONS AND COMPLETE THE CIRCUIT

Once the resin has fully cured, plan your circuit and solder the LED legs in parallel to the battery holder (Figure Ⓔ). The positive and negative legs of the LEDs are uninsulated. To avoid short circuits, use heat-shrink tubing to insulate the LED legs.

FINISHING

Sometimes your resin project has uneven edges — to smooth them, use an electric nail file or fine-grit sandpaper. However, sanding your project can reduce the shine — to give it a crystal shine, you can polish or use doming resin. Your finished resin piece can be further modified by hand-drilling hooks or adding pins to make badges.

Embedding simple electronics like LEDs into resin is just a first step in making wearable fashion technology. With bigger projects, various electronics including sensors can be embedded, making resin a flexible material to work with! ◗

CAUTION: Work in a well-ventilated area, use gloves, and wear a mask while working with resin.

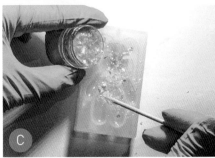

WARNING:
UV resin heats up when curing!

NOTE: Silicone molds that have been used for resin should not be used with food.

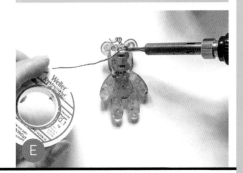

TIME REQUIRED:
A Weekend

DIFFICULTY:
Moderate

COST:
$50–$90

MATERIALS

- » **Clear polymer resin (500g — 250g resin, 250g hardener)** TotalCast
- » **UV resin, hard (200g)** Qiao Qiao DIY
- » **Colorant**
- » **Glitter, confetti, and other embellishments**
- » **LEDs** I prefer color-changing 3mm LEDs for small projects
- » **Coin cell battery, CR1220 or CR2032**
- » **Coin cell battery holder** Kitronik

TOOLS

- » **Measuring cups** for 1:1 measurement of TotalCast resin
- » **Bamboo skewers**
- » **Silicone molds, translucent and opaque**
- » **Long reach lighter or match**
- » **Ultraviolet (UV) light**
- » **Electric nail file or sandpaper**
- » **Soldering iron** temperature controlled
- » **Solder with flux core**
- » **Wire cutters**
- » **Gloves**
- » **Mask**

TIPS:

- » During the curing process, try to keep your project sheltered from dust.
- » When using UV resin, remember to avoid working in natural sunlight, as this will cure your resin.
- » To test whether your project needs more time to dry, use your knuckle to gently tap the surface — to avoid leaving fingerprints.
- » You can diffuse your LEDs' light by sanding the acrylic surface to help spread the light further through your resin. Alternatively, you can buy diffused LEDs.
- » Always test that your electronic components are functional and plan your circuit before embedding it in resin. Once cured, it is not possible to reverse the process.
- » Silicone molds and measuring cups can be washed with soapy warm water. To clean tools that cannot be immersed in water, use kitchen wet wipes, as they contain detergent.

BECKY STERN
is a content creator at Autodesk/Instructables and the author of hundreds of tutorials, from electronics to knitting. Previously, she was a video producer for *Make:* and director of wearable electronics at Adafruit.

KATE HARTMAN
is an associate professor at OCAD University in Toronto where she leads the Social Body Lab, a team dedicated to exploring and developing body-centric technologies in the social context.

'Sup Brows

Written by Kate Hartman and Becky Stern

Ping your bud with the lift of your eyebrows

We spend a lot of time sending messages to each other by texting, emailing, and more. What if you could send a message to a distant friend just by raising your eyebrows?

In this wearable electronics project you'll learn to make your muscles send text messages. The MyoWare muscle sensor uses EMG (electromyography) to sense the electrical activity of your muscles, then converts that into a varying voltage that can be read on the analog input pin of any microcontroller. Here you'll use a Bluefruit

Feather microcontroller to transmit a signal via your phone to two IoT web services — Adafruit IO and If This Then That (IFTTT) — to trigger an SMS text. Let's get started.

BUILD THE CIRCUIT

1. Cut 3 lengths of silicone-covered wire and strip both ends (Figure Ⓐ). Tin the tips with solder to prevent the ends from splaying.

2. Solder one end of each wire to the +, –, and SIG connections on the MyoWare Muscle Sensor (Figure Ⓑ).

TIP: Label the other end of each wire so you'll know which is which. (Figure Ⓒ)

3. Braid the wires (Figures Ⓓ and Ⓔ). This will prevent them from getting snagged when you're wearing the sensor and will create a lovely flexible cord. Tape the board to the table while you work.

Use a zip tie to finish off the end of the braid (Figure Ⓕ). Snip off the tail once it is in place.

4. Position the end of the braid over the top of the Feather board and bring the wires around the edges and up from underneath (Figure **G**). This will ensure that you don't have any scratchy solder connections on the back that will rub against your clothing or skin.

Solder the connections as follows:
» MyoWare + to Feather BAT
» MyoWare – to Feather GND
» MyoWare SIG to Feather A0
Trim the wire ends. Your circuit is complete!

PROGRAM YOUR 'SUP BROWS
Head online to learn.adafruit.com/heybrows/code to install the code on the Feather board, test the Bluetooth link to your phone, create a data feed on Adafruit IO, and create a "recipe" on IFTTT to send a text message (we used Android SMS).

USE IT!
To use your 'Sup Brows, disconnect the Feather from the computer. Connect the battery.

Clean your forehead with some rubbing alcohol to make sure it is free from dirt, oil, makeup, and lotion.

Attach 3 electrodes to their connectors on the muscle sensor: one on the black wire and two on the circuit board. Remove their paper backings, then position the sensor on your forehead as shown: The wires should point up toward the hairline and the sensor should sit at a diagonal, with the lower end above the inside edge of the eyebrow and the higher end shifted away from the center (Figure **H**).

Position the third electrode on your temple. (This one should always be placed away from the muscle that is being sensed.)

Now whenever you raise your eyebrows it will send a "'Sup" message to your friend!

TAKE IT FURTHER
'Sup Brows are just the beginning. Because it's connected to IFTTT, the possibilities of what you can accomplish by just raising your eyebrows are endless!

You can also sense a variety of other muscles to trigger other activities by facial expressions, gestures, and actions. What kinds of networked muscle sensing projects could you imagine? ◐

Get more details on the Adafruit Learning System: learn.adafruit.com/heybrows

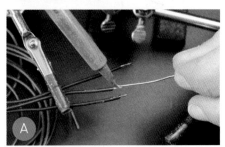

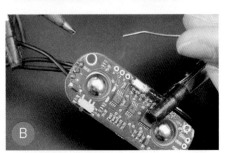

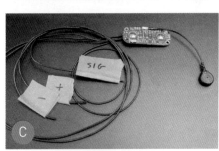

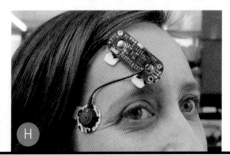

N. Maxwell Lander, Social Body Lab

TIME REQUIRED:
3–4 Hours

DIFFICULTY:
Easy/Moderate

COST:
$80–$100

MATERIALS
» **MyoWare Muscle Sensor** Adafruit Industries #2699, adafruit.com
» **EMG electrodes (3)** Adafruit #2773
» **Adafruit Feather 43u4 Bluefruit LE microcontroller** Adafruit #2829
» **LiPo battery, 3.7V** We used the 500mAh version, Adafruit #1578.
» **Stranded wire, silicone covered** Adafruit #1970. It's strong and very flexible.
» **USB micro cable** Adafruit #2185
» **Cable tie, nylon** aka zip tie

TOOLS
» **Soldering iron and solder**
» **Snips**
» **Masking tape**
» **Helping hands tool**
» **Rubbing alcohol**

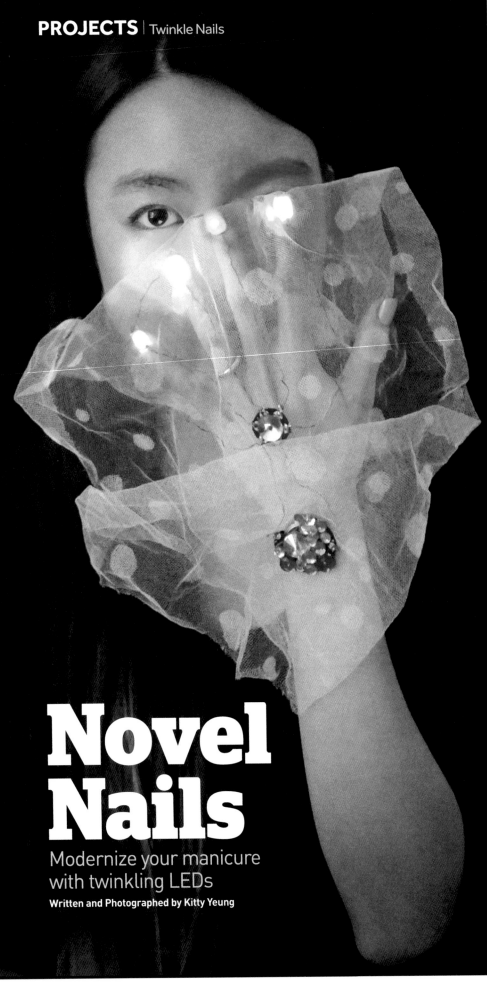

Novel Nails

Modernize your manicure with twinkling LEDs

Written and Photographed by Kitty Yeung

TIME REQUIRED:
2 Hours

DIFFICULTY:
Easy

COST:
$25–$30

MATERIALS

- » **LilyTwinkle microcontroller** SparkFun #11364, sparkfun.com
- » **LilyPad coin cell battery holder, switched, 20mm** SparkFun #13883
- » **LilyPad Rainbow 7-LED strip** SparkFun #13903
- » **LilyPad White 5-LED strip** SparkFun #13902
- » **Conductive thread**
- » **Fabric, tulle or similar**
- » **Battery, coin cell, 3.7V, CR2032**
- » **Gemstones**
- » **Faux nails**

TOOLS

- » **Needle**
- » **Paper**
- » **Hot glue gun**

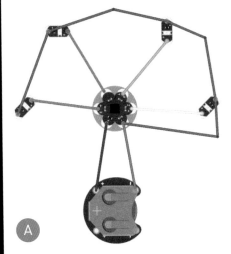

KITTY YEUNG is a physicist, tech-fashion designer, artist, and musician based in Silicon Valley, California. kittyyeung.com

A

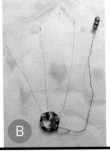

B

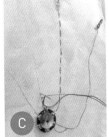

C

About half a year ago, after seeing my tech-fashion designs, a friend sent me a music video with people dressed as robots in light-up dresses, but what really caught my eye was their LED nails. I thought it would be cool to make some.

I wanted them to look natural and not too long or bulky. I also wanted to use microcontrollers to create a twinkling effect. But since fingernails are so small compared to modular electronic components, I had to come up with a solution. What if I made a beautiful bracelet that could hide the electronics inside and connect to the nails?

Here's the outcome.

ASSEMBLE THE CIRCUITRY

For this project, a SparkFun LilyTwinkle, coin cell battery holder, and LilyPad LEDs are used. All these components are thin and sew directly together with conductive thread. Just follow the diagram in Figure A (for beginners, see the example tutorial at makezine.com/go/sparkfun-circuits). All ground lines go to (−).

The nice thing about LilyTwinkle is that it's already pre-programmed to blink the 4 ports randomly. (If you want to reprogram it, check out the tutorial at makezine.com/go/reprogram-lilytwinkle.) I chose to use 4 LEDs exactly as shown in Figure A, so that each LED is on one finger. You can add more LEDs, and of course can share one of the ports to have a fifth trace for the thumb. Always, always test the circuit before sewing any connections.

SEW IN THE COMPONENTS

Now you're good to start sewing the components onto the tulle fabric. Here's a tip: Tulle is flimsy and not very easy to sew thread onto, even if you use an embroidery loop. After sewing the first trace, I realized a better way. You can lay the tulle onto a piece of paper and draw the traces based on the lengths of your fingers (Figure B).

Sew along the traces by poking your needle through both the tulle and the paper (Figure C). The rigidity of the paper gives you a very good grip, allowing definition of the traces. Retest the connections again after sewing.

ADD DECORATIONS

As a maker known for making nerdy stuff pretty, I'm usually not satisfied with bare circuitry. Here I'm pursuing a look where the whole nail lights up, instead of just point light sources, thus I added the faux nails. This is the first time in my life dealing with faux nails (I can't wear them unless I cut them short, as I need to play the piano.) I'm very happy with the strong grip of the glue on their backside. Stick them directly onto the LEDs. The faux nails are thin and diffuse the light so the whole nail glows!

Also notice that I've decorated the LilyPad battery holder and LilyTwinkle with gemstones (Figure D). There are many ways you could decorate these. I just have loads of gemstone leftovers from my tinyTILE dress (hackster.io/kitty-yeung/intel-curie-tinytile-dress-accelerometer-optical-fibers-274294).

THE BIG REVEAL

It is immensely satisfying to peel off the paper from the tulle. On the backside of the tulle, apply double-sided tape to affix it to your hand (Figure E). You can also sew on some ribbons to attach to your hand but double-sided tape is easy, disposable, and re-applicable.

Now put the whole thing on your hand (Figure F). The tulle is soft yet has enough resilience to prevent the conductive threads from crossing and shorting the circuit.

I wish there were other colors of conductive threads, though. I may coat them somehow to make them less visible, or change the color of the tulle, although I really like the white bubbly look. You can also add more LEDs — see Figure G for a circuit diagram.

GOING FURTHER

Making these bracelet/twinkle nails is simple, you just need to find the right "substrate" for the conductive traces. A glove would be an option but I don't recommend it — nails look pretty on fingers. Nails coming out of a glove made of fabric will look creepy. Go for it if it's for Halloween; perhaps a lace glove would work.

You can also try double-sided conductive tape like Alex Glow's LED tattoo (hackster.io/glowascii/glowing-led-tattoo-718a0d).

What I think would be most beautiful is to have chains as the conductive traces. But unless I can find special chains or treatment on the chains, it'd be so hard to avoid crossing (shorting) and to insulate them from the hand. Until then, this remains a romantic idea (Figure H). ⊘

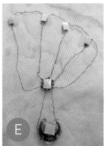

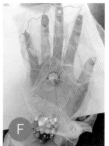

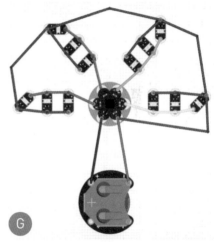

A Good Sign

How I created my makerspace's inexpensive, suspended LED signage Written by Jarrod Hicks

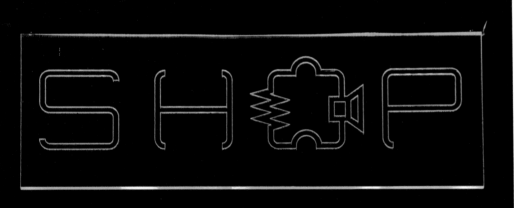

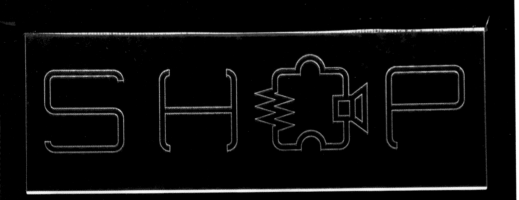

TIME REQUIRED:
4–6 Hours Per Sign

DIFFICULTY:
Moderate

COST:
$25–$70 Per Sign

MATERIALS
» **Acrylic sheet, ⅜"**
» **SMD 5050 LED strip lights**
» **RGB dimmer**
» **RGB amplifiers** amount depends on wire lengths and number of LEDs powered
» **Power supplies, 8.5A 12V, single output switching** Mean Well LPV-100-12, one for the dimmer and each amplifier
» **Stranded speaker wire, 18AWG, 4 conductor**
» **Solid wire, 22 gauge, 4-conductor**
» **Heat-shrink tubing**
» **Duct tape, 2" wide, black**
» **Bridle hooks** aka bridle rings
» **Terminal block connector**
» **Eye lag screws**
» **Hanging wire, 12 gauge**

TOOLS
» **Soldering iron**
» **Laser cutter**
» **Gloves**
» **Wire strippers / cutters**
» **Scissors**
» **Drill/driver**
» **Driver bit for eye lag screws (optional)**

Noisebridge is committed to providing a 24/7 accessible space in San Francisco available to all hackers and makers. They are losing their lease as of August 2018 and are currently fundraising to find a new location. To help out: donate.noisebridge.net

JARROD HICKS
is a maker and designer based in San Francisco, California, who works in the architecture field and volunteers at Noisebridge Hackerspace.

Two things that most any hackerspace can benefit from are better signage explaining the space, and festive LED decorations.

In preparation for Noisebridge's 10-year anniversary in 2017 we decided to address both in one project; creating suspended LED signs that identify the work areas of our space in San Francisco. This type of signage is not uncommon, but our extensive installation and our means of keeping costs down led to several requests for a tutorial. You can read more about it at makezine.com/go/noisebridge-edge-lit-signs, but here's a quick overview of how we made and installed our 14 suspended, edge-lit LED signs.

LINEWORK

The designs of the signs went from sketchbook to AutoCAD to RDworks, which imports .dxf files. I used 3 or 4 line colors: one that was intended to cut through the acrylic, one for cutting halfway through the acrylic, and one that defined the bounds of the sign but was not cut at all (Figure A). On a few signs I used a fourth color to create a shallower hatch across the letters to help them pick up more light. For suspending the sign, I cut holes to pass wire through in the upper corners of the sign. For text, I tried to keep it all around 4" inches tall or taller for easy reading across the room.

CUTTING

I used ⅜"-thick acrylic because it was abundant in the TAP Plastics scrap bin. The speed and power (S/P) settings I arrived at for our cutter were 7/55 cut through, 30/50 for a ~3/16" deep cut.

The half-deep cut forms the graphics of the sign while the cut-through in the upper corners is used for mounting holes. To protect the signs from all the handling, I kept most of the protective film on them until they were mounted and tested.

ELECTRONICS

I determined that ESP8266/Arduino boards with fancy LEDs would be too expensive and ended up using an RGB dimmer (Figure B) to control the whole thing. This also meant using SMD 5050 strip lights, which are cheap. It was also easier for me to set up, as I am still new to electronics.

This budgetary decision turned out to be an improvement for the project overall. These dials sit next to the upstairs entrance to Noisebridge and allow anyone to change the color of the signs in the space. Figure C is the sketch of the layout of the system.

For the RGB dimmer to run all 14 signs, I split the system so that the signs on the west and east sides would run off of separate amplifiers (Figure D), which I installed on the walls sort of centrally located to the signs they would serve.

While the dimmer, LEDs, and amplifiers were cheap, I opted to get reliable Mean Well LPV-100-12 8.5A 12V single output switching power supplies. I used 18AWG 4-conductor stranded speaker wire between the dimmer, amplifiers, and signs, run along the ceiling using bridle hooks. I went with cheap, easy to use terminal block connectors (Figure E) on the ceiling above where each sign would be, allowing me to detach and reattach each sign's LEDs with only a screwdriver.

To suspend the signs I used eye lag screws in the ceiling, with a special driver bit made for them, and hooked 3' lengths of 12 gauge hanging wire onto them — sturdier than I needed, but it allowed me to adjust the signs' positions by simply bending the wire.

SIGN ASSEMBLY

I cut the LED strips to the length of the signs, then attached enough 22 gauge 4-conductor solid wire to reach the ceiling when the sign was installed. I had to be very careful bending and moving these connections (Figure F), as they are extremely fragile, even with heat-shrink tubing applied.

To keep costs down, the LED strips were attached to the tops of the signs with 2"-wide black duct tape (Figure G). I folded any extra tape over to close off the light at the end and trimmed the excess with scissors. A very small amount of light bleeds through, but not enough to require more tape or another solution. Ideally each LED is centered facing the edge of the sign, but a little offset doesn't seem to make a big difference.

I hope my notes can help you make your own (improved!) version for your space. ⊘

A

B

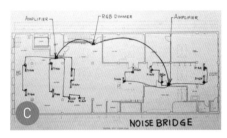

C

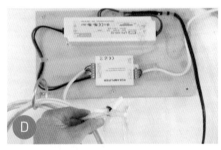

D

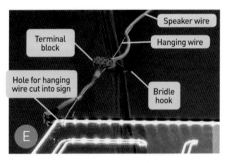

Terminal block
Speaker wire
Hanging wire
Hole for hanging wire cut into sign
Bridle hook

E

F

G

Power Ranger

Get notified by SMS when the electricity goes out

Written and photographed by Mari DeGrazia

MARI DeGRAZIA (@maridegrazia) is a director at Kroll Cyber Security specializing in digital forensics and incident response. She enjoys destroying her nails by working on projects.

My father needed a way to monitor power and temperature at a remote location in Alaska while he was out of state. When the power went out or the temperature dropped, he wanted a notification. Since Wi-Fi wouldn't be available during a power outage, the notification method I thought made the most sense was SMS. My remote monitor would also need to run from a battery for several hours. Based on these requirements, I decided to use the Adafruit Feather Fona with an HTU21D-F temperature sensor.

The Feather Fona uses a 2G SIM card to send and receive text messages. It can be powered from a lithium battery, which is recharged directly from the Fona. A voltage reading can be taken from the Fona's USB pin, which makes it perfect for detecting an outage: If the voltage is low, this means it's running from the battery, which means the power is out.

I added an LED Matrix to display status messages, and a prototyping board to build the voltage and temperature circuit. By using stacking headers, I was able to fit all the components nicely in a 3D printed case.

1. PRINT THE CASE (OPTIONAL)

Not strictly necessary for this project, my 3D printed case (thingiverse.com/thing:2758688) is remixed from the Adafruit Feather case.

2. ASSEMBLE THE FEATHER FONA

Follow the instructions that came with the SIM card to activate it, then insert it into the Fona. Solder the stacking pins with the female headers on top, and the male headers on the bottom (Figure Ⓐ).

3. ASSEMBLE THE LED MATRIX

Follow the instructions at learn.adafruit.com/adafruit-8x16-led-matrix-featherwing/assembly to build the matrix (Figure Ⓑ).

4. BUILD THE VOLTAGE CIRCUIT

The Fona's USB pin is the positive voltage from the USB jack. The battery connects to that pin through a diode that cuts the battery off when USB power is present. To determine when the power is out and just the battery is running, a voltage divider is needed. Thanks to Adafruit's Mike Stone for the tips on how to do this. On the prototype board, use two 100K resistors to build one between the USB and GND pins. Connect the middle of the divider to pin A5, as shown in Figure Ⓒ.

5. WIRE THE PROTOTYPE BOARD

For the temperature sensor, connect one wire each to the SDA, SCL, power, and ground holes on the prototype board (Figure Ⓓ). For the slide power switch, connect one wire each to the EN hole and a ground hole (Figure Ⓔ). Now solder the female headers onto the prototype board (Figure Ⓕ).

Don't connect the sensor or switch at this time, since the wires will need to be threaded through the case in a later step.

6. FINAL ASSEMBLY

Mount the proto board to the bottom of

TIME REQUIRED:
3–4 Hours

DIFFICULTY:
Moderate

COST:
$100–$120

MATERIALS
» **Adafruit Feather 32u4 Fona microcontroller, with GSM uFL antenna** Adafruit #3027 and 1991, adafruit.com
» **2G SIM card** Adafruit #2505
» **FeatherWing prototyping board, with female and stacking headers** Adafruit #2884, 2886, and 2830
» **HTU21D-F Temperature & Humidity Sensor board** Adafruit #1899
» **8×16 LED Matrix Display** Adafruit #3155
» **Slide switch, SPDT** Adafruit #805
» **Lithium battery, 3.7V, 500mAh or higher**
» **100K resistors (2)**
» **Hookup wire**
» **M2.5 spacers (4)** such as Amazon #B01N9Q8YLE
» **3D-printed case (optional)**

TOOLS
» **Soldering iron and solder**
» **Phillips screwdriver**

the case using the M2.5 spacers. Feed the sensor wires out the hole in the side, and the switch wires though the square hole in the bottom (Figure G). Now solder them.

Place the switch and battery in the battery case, and snap the battery holder onto the bottom of the case (Figure H). Plug the Fona into the prototype board and thread the antenna through the antenna hole (Figure I). Plug the battery and the LED matrix into the Fona (Figure J).

7. CONFIGURE THE CODE
First, read Adafruit's tutorial to make sure your Fona is up and running: learn.adafruit.com/adafruit-feather-32u4-fona/setup.

In the Arduino IDE, go to Sketch → Include Library → Manage Libraries, then type the following library names in the filter box and select Install for each one:
» *Adafruit HTU21DF_Library*
» *Adafruit FONA*
» *Adafruit GFX*
» *Adafruit LEDBackpack*

Download the power monitor code from github.com/mdegrazia/PowerMonitor. Open the *remote_power_monitor.ino* file and look for the section titled **USER CONFIG**. Change the phone number to the one you want to receive notifications, and adjust the low temperature to the desired values. Upload to the Fona.

USING YOUR POWER MONITOR
Place it in a location that has strong cellphone reception and plug it in. (A stronger antenna can be used if the signal is weak.) Status messages will display on the LED matrix indicating that it has connected to the GSM network, the current signal strength, and battery level.

It will send a push text notification for the following events:
» When the device first powers on and starts monitoring
» When the power is out
» When the power comes back on
» When the temperature is low

To get a status update, send a text message that says "Status" to your Remote Power Monitor. The monitor will reply with the current power status, battery level, voltage, and temperature.

If the Power Monitor is not "tripping" when the power goes out, the voltage level threshold may need to be adjusted. Open the Arduino Serial Monitor while your Power Monitor is plugged in and note the voltage levels. Unplug the Power Monitor, send several "Status" text messages, and note the voltage levels. The voltage should be lower when the power is unplugged. Update the **voltageThreshold** value in the **USER CONFIG** section accordingly.

GOING FURTHER
Next, I plan on incorporating an optional MQTT feed over GPRS which will allow constant monitoring of power, battery level, and temperature, all from the phone. Gas or motion sensors could also be added to the Power Monitor to enhance its capabilities. ✪

Written by William Gurstelle

Gimme Shelter

Build a model pyramid and learn how the Mayans used lime to make monuments lasting millennia

WILLIAM GURSTELLE's new book series *Remaking History*, based on this magazine column, is available in the Maker Shed, makershed.com.

When ancient peoples first began building things, they did so by piling one rock upon another to make walls. Of course, those rocks were neither flat nor square, so there were gaps between them. As far back as 6500 BCE, people used crushed rock and water to make a paste to fill the gaps. If the right crushed rocks were used, then the paste would harden when it dried, holding rocks together, making the wall impervious to sun, wind, and rain. Such a paste is called *mortar*.

Early on, primitive experimenters found that only a few types of crushed rock would harden and bond together permanently. The best ancient mortars were made from crushed and heated limestone and sand. Homes, palaces, temples, and so on made with limestone-based mortar would harden into rock-stable structures that could last lifetimes. This was some pretty great stuff, in fact, so great that many historians feel that *lime mortar*, which ancient builders in both the Old World and the New World discovered independently, is one of the most important chemical discoveries in the history of civilization.

What exactly is lime? The name is a bit confusing because chemical lime has nothing to do whatsoever with the tart-tasting fruit found in lemon-lime soda or key lime pie. Chemical lime is what you get when you crush limestone (which is mostly calcium carbonate) and then heat it up to high temperature. At that point the limestone releases carbon dioxide and turns into calcium oxide, which is known as *quicklime*. Quicklime is fairly nasty stuff because it is highly caustic and has the bad habit of reacting violently with water, creating copious amounts of heat. But after water has been added, or "slaked" as old-time masons would term it, the resulting chemical compound, calcium hydroxide, also called *slaked lime* or *hydrated lime*, is much more stable and less corrosive.

Why is lime such an important chemical? Because it has been used by so many peoples and for so many different purposes. Lime has been, and still is, used as fertilizer, for purifying water, for making glass, smelting metal, and above all, for cementing rocks to one another.

In fact, for the people of Mesoamerica, no chemical was as important as lime, for not only did they use lime to make mortar

to build the huge temples of the Mayan and Aztec civilizations, it was also the key to their maize-based diet.

The ancient Mesoamericans used lime to prepare their food. The use of chemical lime to process maize is called *nixtamalization*, and was known to exist as early as 1000 BCE. The ancient Aztec and Maya mixed lime with water to make calcium hydroxide, which was known by the Spanish as *cal*.

Cal was a critically important part of what the Mesoamericans ate because untreated corn can't provide the vitamin niacin. Cooking maize with lime frees its niacin so it can be absorbed by the body. Combined with beans (for amino acids), nixtamalized maize helped provide a balanced diet. Even today, Latin American tortilla makers use significant amounts of cal to make tortillas and other regional food staples.

Another important use of cal was to make the mortar that holds together the stones of the huge temples at places like Teotihuacan, Tikal, and Chichen Itza. This simple but vital building material is strong and tough enough to last a thousand years.

A MAYAN STEP PYRAMID
If you visit the famous Mayan ruin called Chichen Itza in Mexico's Yucatan peninsula, you'll see a giant stone rendering of the feathered serpent god Kukulkan at the base of the four-sided step pyramid called El Castillo. El Castillo is 98 feet tall and 181 feet wide, and is constructed of huge limestone blocks cemented together with lime mortar. In ancient times, architects also used the lime to make a stucco to finish the exterior of temples and pyramids.

In this edition of Remaking History we explore how the lime was used to construct buildings in ancient times by making a model Mayan-style structure inspired by the great Temple of Kukulkan.

BUILDING YOUR PYRAMID
1. CONSTRUCT THE SUPERSTRUCTURE
Begin by nailing the 14" 2×4s to the plywood base to form a square, as shown in Figure ⑧ on the following page (I built mine with a cutaway corner for showing the pyramid's interior). Then, complete the pyramid's superstructure by placing and nailing the 10½", 7", and 3½" 2×4s atop each other.

f9photos / Adobe Stock, William Gurstelle

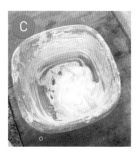

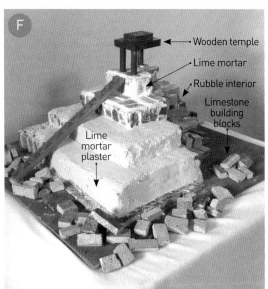

Wooden temple
Lime mortar
Rubble interior
Limestone building blocks
Lime mortar plaster

2. PREPARE THE LIME MORTAR

Put on your safety glasses and rubber gloves. Mix mortar in small batches by placing 5oz of cal, 10oz of sand, and a small amount of water in a plastic pail. Thoroughly mix the ingredients, and then carefully add enough water to form a medium-thick paste (Figure C).

3. TILE THE FIRST WALL

Begin adding stones to the vertical walls of the base layer of your pyramid. Using a small spreading knife or spatula, evenly apply a thin coat of mortar to the mating surface of the stone, then press the stone firmly against the side of the wood superstructure. Continue to add mortared stones all around the base, taking care to keep the mortar lines even and straight.

4. TILE THE FIRST STEP

Once the sides of the base are completely tiled, add mortared stones to the horizontal surface of the base (Figure D).

5. REPEAT

Then do the next wall, and the next step. Continue to add stones and mortar to the pyramid, one step at a time, until the surface is fully tiled.

6. LET IT DRY

When all the blocks are installed, let the pyramid lie undisturbed while the mortar dries (Figure E). Note that it takes weeks for lime mortar to cure completely. Uncured, it has little strength. If mortar cracks appear, reapply mortar as needed.

7. PLASTER THE EXTERIOR

Once partially cured, plaster the exterior by mixing another batch of lime mortar. Add enough water to make it medium-thin consistency and apply with a spatula or a narrow, drywall taping knife.

8. ADD DETAILS (OPTIONAL)

Unlike the pointed pyramidions that cap Egyptian pyramids, Mesoamerican pyramids have a broad, flat top. Typically, a brightly colored wooden temple sat atop the pyramid, which was used by priests as a place for religious ceremonies. You can add authenticity to your model by building a temple structure or staircase from dowels and craft wood (Figure F). ✎

CUTTING STONES

Mesoamerican pyramids were typically built from cut limestone blocks. The blocks used could weigh many tons and were hauled into place through the use of levers, ramps, and a great deal of human muscle power.

You can cut limestone blocks to size using the ancient methods of the Mayans, but I wouldn't recommend it. Limestone is very hard, and even with appropriate chisels and hammers, cutting rocks to size and shape would take a very long time.

For this model pyramid, you can use a tile saw to cut unglazed limestone floor tiles (or any other type if you can't find limestone tile) to about 1"×2". If you don't have a tile saw, you can rent one inexpensively, or alternatively, look in the tile section of large home stores for 12"×12" mosaic tiles which consist of approximately 1"×2" rectangular stones cemented to a flexible mat. It's easy to remove the smaller stones from the flexible mat for use in this project.

William Gurstelle

1+2+3 Simple Succulents

Written by Jennifer Refat

Brighten your home with these easy to make, zero-maintenance paper plants.

1. PRINT AND CUT PLANT PIECES

You'll need 30 leaves in 3 sizes: 10 large, 10 medium, and 10 small. You'll also need one 1½" circle for the plant base. Print and cut out the leaves from makezine.com/go/paper-succulents-template. For all large and medium leaves, cut a 1" slit at the bottom.

2. GLUE AND FOLD LEAVES

For all 10 large and 10 medium leaves, overlap the two slits on each and glue the overlap using a glue stick. The overlap should be slight. This will help prop them up and give them a natural shape.

For the small leaves, fold up ¼" of the bottom. This bottom fold will help you place the leaves down when gluing them.

3. ASSEMBLE A PLANT

Use the glue gun to place 5 leaves equidistant on the circle's outer edge. Repeat with 5 leaves at a time for each layer, from largest to smallest, working your way into the center. Alternate the layers so that you're placing new leaves between the previous leaves. You will have 6 layers.

Using a pencil, curl the outermost leaves outward and the rest inward.

Make several succulents and place in clay pots for a pleasing paper garden. ⊘

TIME REQUIRED:
1 Hour

DIFFICULTY:
Easy

COST:
$1–$3

MATERIALS
» Cardstock paper (2 sheets)
» Leaves template
» Small clay pot (optional)

TOOLS
» Scissors
» Glue stick
» Glue gun
» Pencil

JENNIFER REFAT
is a software developer who's been crafting since childhood. Originally from NYC and now based in Toronto, she's the founder and creator of Craftic (craftic.com and @crafticland), a DIY website with tutorials inspired by your craft supplies.

Hep Svadja

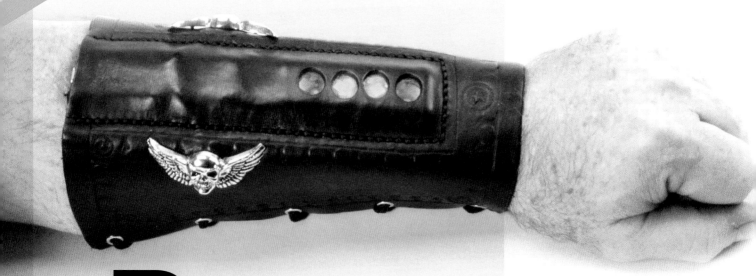

Brace Yourself

Written and photographed by Tim Deagan

Learn about melding electronics and leather with this **light-up project**

MATERIALS

- » Leather, veg tan (3–4oz)
- » Leather stitching thread (25 yards) such as Tandy Leather's Waxed Nylon Thread, tandyleather.com
- » All-in-one dye and finish Eco-Flo
- » Eyelets
- » Conchos
- » Lacing or shoelace
- » Plastic wrap
- » Masking tape
- » Microcontroller I used the Gemma V2, Adafruit #1222 adafruit.com
- » USB LiPoly charger
- » Battery, LiPo, 3.7V, 150mAh Adafruit #1317
- » Flora RGB Smart NeoPixels, version 2 (pack of 4) Adafruit #1260
- » Wire
- » Lead-free solder
- » PLA filament (optional)

TOOLS

- » Cutting knife
- » Straightedge
- » Stitching hole punch
- » Rawhide mallet
- » Hole punches
- » Leather stamps
- » Eyelet setting tool
- » French edging tool #2
- » Saddle stitching needles
- » Soldering iron
- » Multimeter
- » CAD design software Fusion 360, Tinkercad, etc.
- » 3D printer and software (optional) Or you can send the files to a service for printing. See makezine.com/where-to-get-digital-fabrication-tool-access to find a printer or service.

WEARABLE MICROCONTROLLERS HAVE FOUND THEIR WAY INTO COSPLAY, FASHION, AND DAILY WEAR but are usually textile based. In this Skill Builder we'll explore techniques for including leather in your wearables by examining my forearm bracer project, which uses a built-in Adafruit Gemma and RGB NeoPixels that change color patterns with a simple capacitive touch of the medallion on its side. Read along and learn some key tips to working with leather and electronics, both on their own, and combined into one project.

Leather is a favorite material of crafters, cosplayers, and makers for good reasons. It is moldable, cuttable, colorable, and durable. It can be riveted, stitched, glued, and laced. As a substrate for a wearable project it has some very desirable qualities and a few gotchas. The particulars will vary with different efforts, but the general themes are relevant across most projects.

CIRCUIT ACCESS

With any wearables project, but especially one based in leather, how to access the electronics is a key planning decision. When it's time to recharge the battery or reprogram the microcontroller, will you want to remove the items or deal with them in place? A design that keeps things internal typically allows for less wear and tear on the wiring because you can firmly secure and protect the components. Easily accessed or removable components generally make reuse, reprogramming, and recharging a much simpler option.

For me, recharging occurs much more often than reprogramming. So my bracer project provides in-place charging without removing the battery or unplugging it from the microcontroller. Reprogramming does require removing the circuit from the leather to get access to the USB port, but is still possible with the approach I'm taking.

The components I chose to include are a charger board, the Gemma M0, a LiPoly battery and four Flora RGB LEDs (Figure A)

I also wanted to take advantage of a feature of the Gemma that has interesting opportunities when working with leather: capacitive input. Three pads on the Gemma allow sensing a touch to the pad, or something wired to the pad. This is a single wire input (no ground connection).

Leather projects often include rivets, snaps, and decorative **conchos** (metal ornaments) that are all potential candidates for discrete capacitive inputs. I used two conchos that will be wired to the Gemma as a trigger for the lights. Test before you commit to using an object — it's difficult to tell which may be conductive without an ohm-meter or trying them in your circuit.

SELECTING A LEATHER

Leather comes in different thicknesses and types. Most leatherworkers tend to use vegetable-tanned ("veg tan") leather so that they can stamp and dye their projects. "Chrome tan" leather more closely resembles cloth, so I chose veg tan, which I dyed with an all-in-one dye and acrylic finish once constructed. Experienced wearable makers will recognize these issues; they are all present in textile-based projects but are often exacerbated by the differences when working with leather.

MAKE A SABOT

Because leather can abrade wires and solder points, I've become enthusiastic about 3D printing carriers for components that I can insert under or into the leather. Protective outer shells to ease fitting are sometimes referred to as **sabots**. The sabot I designed and printed holds my components (Figure B). I made models of each of them to allow practice fitting in Fusion 360. If you don't have a 3D printer, you could alternatively handcraft a sabot out

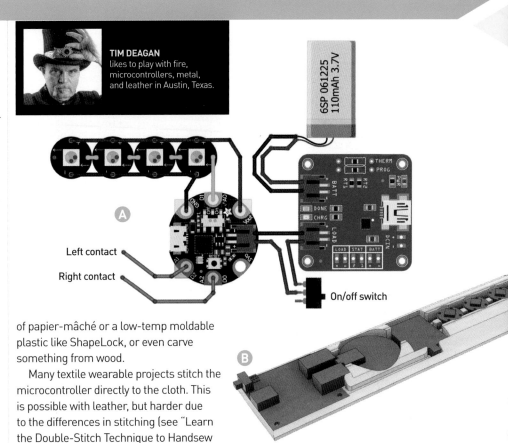

TIM DEAGAN likes to play with fire, microcontrollers, metal, and leather in Austin, Texas.

Left contact

Right contact

On/off switch

of papier-mâché or a low-temp moldable plastic like ShapeLock, or even carve something from wood.

Many textile wearable projects stitch the microcontroller directly to the cloth. This is possible with leather, but harder due to the differences in stitching (see "Learn the Double-Stitch Technique to Handsew Leather" at makezine.com/2017/01/23/handstitch-leather). I've found that making pockets to hold my components, especially when encased in a sabot, is the most durable and flexible approach. I'm combining a molded piece of leather on one side of the pocket and a flat piece of leather on the other. I'll stitch around them to form the pocket for the electronics.

CREATE YOUR PATTERN

Leather bracers are surprisingly easy to make. Typically they're a single piece of leather, but they need to taper to fit the forearm. There are many ways to make a pattern for cutting leather; I like the simple method of wrapping the target area in plastic wrap then covering that with masking tape. I use a marker to identify the shape to cut and where the components are going to be placed (Figure C).

After I cut the tape off and trim it to its marked boundaries, I can place my 3D printed sabot to make sure everything is sized properly (Figure D). I use a ball tip stylus to transfer the pattern to leather and make the desired cuts (Figure E). Don't worry if the pattern won't lay perfectly

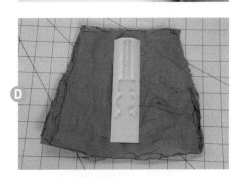

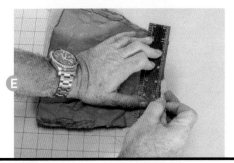

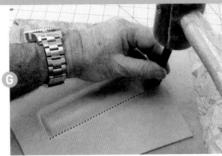

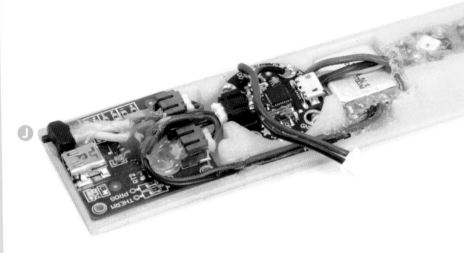

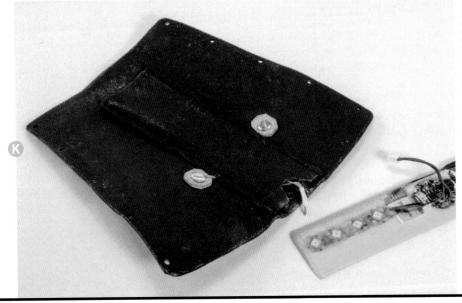

flat. Unlike cloth that needs darts or other techniques to create contours, we can stretch and mold leather.

SHAPE AND STITCH

Leather is easy to mold. I used my 3D print as the model to mold over, and I cut a second pattern to be sure I kept reasonably close to the original outline of the bracer while stretching the leather over the model. Soak your leather (I used 3-4oz.,) in water and place it over the model, stretching it down around the edges with your finger or a blunt object (Figure F). If you leave "scratches," you can rub them out with your finger.

Once the leather dries, it will harden and serve as a protective shell around the model. Be careful when molding items, as leather will shrink a bit when drying and then, with time and wear, generally loosen up again. The challenge is that it can be very difficult to get your sabot in and out of the bracer if the fit is too tight. You can re-wet and stretch the leather or better yet, print a version of your sabot that's 5–10% larger than the one you'll use with components, and use that to mold the leather.

Stitching leather is a bit different than cloth in that you pre-punch the holes. Fork-style punches are an easy way to make consistent holes. I placed a second piece of leather below the top piece and punched my holes through both (Figure G). If you didn't mold using an oversized model, be sure to give yourself a few extra millimeters of slack in how close you make the stitching holes to the model.

Trim the bottom pocket cover to about ¼" around the stitching holes (Figure H). To keep things simple, I used laces rather than buckles or snaps to hold the bracer on my arm. Using leather punches, I made holes for the lacing eyelets, conchos, and the Flora RGB LEDs. I also did some light decorative tooling on the top piece, then dyed and finished the leather (Figure I).

CREATE YOUR CIRCUIT

On my bracer, the assembly of the electronics was straightforward. I placed the components into the sabot and measured out my wiring so that it would fit. After soldering, with careful checks that everything would continue to fit, I used hot melt glue to hold everything in place. To assist in diffusing the LEDs, I covered them in hot melt. This also serves to protect them since they're accessible through the holes in the bracer. Be careful to keep everything within the profile of the sabot so that it will slide in and out of the pocket (Figure J).

The most difficult part of the circuit is the connection to the capacitive inputs. I used screw-back conchos with the wire wrapped multiple times around the shaft. When stitching the pocket, I made sure to bring these in from each side, terminating them in a plug that I could attach to the Gemma (Figure K).

The screw on the back of the concho needs to be isolated from your skin so that it isn't constantly triggered. This could have been done by leaving side flaps on the bottom pocket piece, but I ended up using electrical tape so that I could get at the screws if the wire broke or I wanted to change conchos.

PROGRAMMING

I'm a huge fan of CircuitPython and used a mashup of Adafruit's Flora and capacitive input examples for my initial software. I added a simple "Knight Rider" display that changes colors when one of the conchos is touched and displays the rainbow cycling example when the other concho is touched. You can access this code (Figure L) at makezine.com/go/bracer-code.

While I'd hesitate to use the capacitive input for anything critical (it's a little finicky), I am delighted at how well it works for this project (Figure M).

GO NUTS!

Wearable projects span the gamut from exposed components that look like digital jewelry, to lights and sensors hidden in traditional clothing, to 3D printed crowns and tiaras. Makers have only scratched the surface on this wonderful approach to personal digital wear. I am of the firm position that leather is a fantastic addition to the wearable maker's toolkit. It's protective, it can be dyed colors and beautifully tooled, it's easy to mold and stitch. With the addition of 3D printing and innovative input/output, we are likely to see an explosion of leather wearable projects in the years to come. ⊘

```python
import board
import neopixel
import time

touch0 = touchio.TouchIn(board.A1)
touch1 = touchio.TouchIn(board.A2)
pixpin = board.D1
numpix = 4
strip = neopixel.NeoPixel(pixpin, numpix, brightness=0.3, auto_write=False)

def wheel(pos):
    # Input a value 0 to 255 to get a color value.
    # The colours are a transition r - g - b - back to r.
    if (pos < 0) or (pos > 255):
        return (0, 0, 0)
    if (pos < 85):
        return (int(pos * 3), int(255 - (pos*3)), 0)
    elif (pos < 170):
        pos -= 85
        return (int(255 - pos*3), 0, int(pos*3))
    else:
        pos -= 170
        return (0, int(pos*3), int(255 - pos*3))

def rainbow_cycle(wait):
    for j in range(255):
        for i in range(len(strip)):
            idx = int ((i * 256 / len(strip)) + j)
            strip[i] = wheel(idx & 255)
        strip.write()
        time.sleep(wait)

def nrider_cycle(wait,r,g,b):
    j =len(strip)-1
    while True:
        for i in range(len(strip)):
            strip.fill((0, 0, 0))
            strip[i] = (r,g,b)
            strip.write()
            time.sleep(wait)
        for i in range(len(strip)):
            strip.fill((0, 0, 0))
            strip[j-i] = (r,g,b)
            strip.write()
            time.sleep(wait)
        if touch0.value:
            return 0
            break
        if touch1.value:
            return 1
            break

while True:
    touch_rtn = nrider_cycle(0.05, 255,   0,   0)
    if touch_rtn:
        touch_rtn = nrider_cycle(0.05,   0, 255,   0)
    if touch_rtn:
        touch_rtn = nrider_cycle(0.05,   0,   0, 255)

    if not touch_rtn:
        while touch0.value:
            rainbow_cycle(0.001)      #rainbowcycle with 1ms delay per step
            if touch1.value:
                break
```

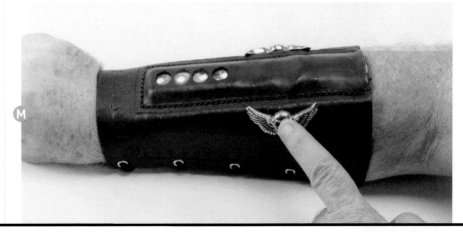

Meddling with METAL

Customize your work with a variety of finishing techniques

Written by Hep Svadja

HEP SVADJA is *Make:*'s photographer and photo editor. In her spare time she is a space enthusiast, metal fabricator, and Godzilla fangirl.

THE FINISH YOU CHOOSE FOR YOUR METALWORKING PROJECT can influence the final look as much as the material or shape can, and there are many beautiful options to choose from.

Your metal surface should be as clean as possible before applying your finish. Muriatic acid is good for removing metal particles and limescale, and industrial metal degreaser removes oils. Pumice pastes can remove grime, especially on small pieces. You may need to roughen your surface for some finishes, using a wire brush or sandblasting before cleaning and degreasing.

FINISHES

❶ PATINAS

To add a patina to silver, steel, bronze, or copper, you can use Liver of Sulfur (LOS). Make an LOS solution and a neutralizing bath with hot water and baking soda (a couple tablespoons per quart).

Submerge your metal in the solution for a few minutes, then dunk it into the neutralizing bath to stop the reaction. Repeat this to further darken the metal. Use a soft brass brush to burnish the finish or even it out.

LOS doesn't work as well with gold or bronze, but you can pickle the piece and plate it with iron to get a patina. Dunk your piece into pickle solution, take it out, and coat it in iron filings, or blot with medium steel wool. Use the neutralizing bath to stop the process.

NOTE: Pickle solution acts as an oxidation agent. Make pickle solution by heating a cup of white vinegar, a tablespoon of salt, and a teaspoon of hydrogen peroxide in a double boiler until not quite boiling. Store and prep in glass or ceramic — this solution will eat metal!

❷ COPPER VERDIGRIS

Mix a solution of 2 parts white vinegar, 1½ parts non-detergent ammonia, and ½ part salt. Use a misting bottle to evenly spray your piece. After an hour, re-mist the piece completely and evenly. Let it sit overnight.

Add more salt to the solution for brighter green, or less for a more gray finish.

③ BLUING STEEL

Hot salts bluing and rust bluing are the most common techniques for this method. Keep materials far away from other metals, as the fumes can cause them to rust.

For hot salts bluing you'll need a black oxide solution. Heat the solution, then immerse the metal in it for 15–30 minutes, making sure the piece is evenly coated. Rinse, scrub with cold water, then place in boiling water for 5–30 minutes depending on size. Put the piece in water-displacing oil overnight.

For rust bluing, polish the piece as much as possible. Apply a solution of 1 part nitric acid and 3 parts water. Let the piece sit until a very fine layer of red rust appears (usually within 24 hours). Once coated, boil the piece in distilled water for 30 minutes. The red oxide should turn dark and appear like black velvet. Remove the oxide with degreased fine steel wool, and repeat as desired. Soak in water-displacing oil for 24 hours.

④ FLAME COLORING

Copper and steel can be brilliantly colored by applying heat with an oxyacetylene torch. Experiment with the intensity and duration to achieve different colors. Keep a container of oil nearby to quench the metal once you're done.

⑤ RUST

Rust on steel is easy to achieve quickly using hydrogen peroxide. Mix together, in order, 16oz of hydrogen peroxide, 2oz vinegar, and ½ tablespoon salt. For an uneven coating, apply the rust solution straight to the clean piece, repeating as desired. For a more uniform distribution, apply pickle solution to etch the metal and let it dry. Repeat this several times, then apply rust solution.

SEALING

Once you've finished your metal, sealing it with a protective layer of wax, oil, or lacquer will protect it from oxidation. Most sealants, especially wax, need to be reapplied periodically — just clean and dry your piece, then reapply the sealant.

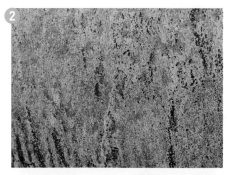

WAX

Wax is especially good for preserving a dull, matte, or textured finish. You can use almost any kind. Automotive waxes sometimes have additives for increased shine, which can really bring out polish or flame color finishes. Heat your piece just enough so that the wax can evenly adhere to the surface and caramelize, but not so hot that it burns off or the metal discolors. Apply wax with a brush, or melt it directly onto the metal, and use a soft paintbrush to spread it around while hot. Once the wax stops bubbling, quench the piece in water, then buff with a soft cloth.

BOILED LINSEED SOLUTION

Boil a solution of 3 parts linseed oil and 2 parts turpentine with a dash of any oil-drying agent. Apply in a thin coat with a paintbrush, and rub off the excess. This method takes about a week to fully cure.

POLYURETHANE AND LACQUER

You can rub lacquer directly into dry metal in thin coats with 0000 steel wool. Finish with paste wax and buffing. Use a spray can to apply polyurethane directly to dry metal. In a draft-free but ventilated area, spray 8" above the surface in smooth, same direction strokes. Make sure to coat completely and evenly. Wait 15 minutes, then reapply two or three more coats. ◆

shotsstudio / Adobe Stock, donatas1205 / Adobe Stock, jbphotographyt / Adobe Stock, lumikk555 / Adobe Stock, mrkyle229 / Flickr, romantsubin / Adobe Stock

CAUTION: Many of these methods produce toxic fumes and corrosive chemicals. Wear personal protective equipment, and work in a well-ventilated area. Use tongs as a buffer. Don't use metal containers or utensils — they'll affect the finish of your piece. Opt for glass, ceramic, or plastic.

Fab FILLETS

*Get up to speed on designing the most common **joints for CNC** construction*

Written by Anne Filson, Gary Rohrbacher, and Anna Kaziunas France

ANNE FILSON
is an architect, educator, and co-founder of the architecture, design, and research firm Filson and Rohrbacher.

GARY ROHRBACHER
is an architect, professor, and co-founder of Filson and Rohrbacher.

ANNA KAZIUNAS FRANCE
is the co-author of *Getting Started with MakerBot* and compiled *Make: 3D Printing*

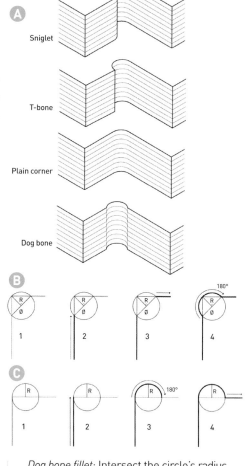

*The following is an excerpt from our new book **Design for CNC**, available now.*

A **fillet** is a design feature that rounds off a corner. When it comes to cutting flat parts, it's imperative to add fillets to the interior corners if you want your joinery to fit properly.

Two commonly used fillet solutions are **dog bones** and **T-bones** (Figure Ⓐ), named for their appearance. When a dog bone fillet is added to both sides of a slot, the slot takes on a cartoonish, dog bone- or T-bone-like shape.

EXERCISE:
How to draw fillets

The best way to understand how fillets work and why they look the way they do is by drawing them yourself. Plus, it's great practice for creating or adapting your own designs. Fire up your favorite vector graphics-capable program (something like SketchUp, Inkscape, Illustrator, VCarve Pro, or AutoCAD) and follow along.

Dog bones and T-Bones

Dog bone and T-bone fillets are very similar; the core difference is where the "circle" that will accommodate the tool diameter is placed (relative to the inside corner you're trying to eliminate).

Choose your tool diameter, or Ø. A ¼" diameter tool is most commonly used for large but detailed CNC projects. It's strong enough to withstand the cutting forces and long enough to pass through ¾"sheet materials without breaking.

Create a circle that is 110% larger than the tool's diameter. For example, if you were using a ¼" diameter tool, you'd draw a 0.275" (7mm) diameter circle with a 0.1375" (3.5mm) radius.

Place the circle over the inside corner. This is where dog bones and T-bones differ. The dog bone fillet rounds out the corner, while the T-bone pulls the circle out to the side of the vertex.

Dog bone fillet: Intersect the circle's radius, R, with the inside angle's vertex and its diameter, Ø, with the part edges (Figure Ⓑ).

T-bone fillet: Align the circle's diameter, Ø, with the inside angle's vertex along one side of the part (Figure Ⓒ).

Integrate the circle into the part lines. As shown, draw a stroke, create a line, or use a Boolean operation to assimilate the circle into the overall part. Ensure that the lines are connected.

NOTE: Fillets are tool diameter dependent; their size and shape are determined by the diameter of the end mill used to cut the parts.

Because you're hardcoding the tool size into your design, it's a good practice to draw fillets slightly bigger than necessary — 110% larger than the actual tool diameter works well. This ensures that if the drawing needs to be scaled down slightly, the end mill will still fit onto the inside corners. ◗

Filson and Rohrbacher

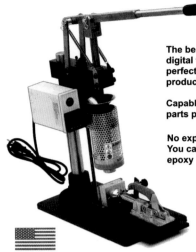

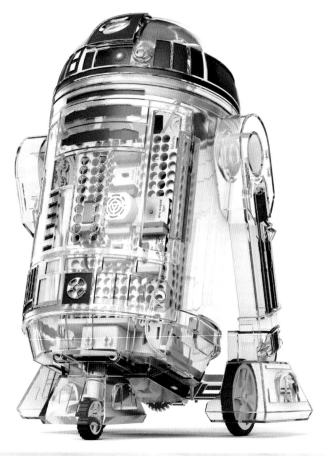

LITTLEBITS DROID INVENTOR KIT

$100 shop.littlebits.cc

I've always loved the littleBits sets for younger makers, so I was eager to get my hands on the new littleBits Droid Inventor Kit to see how it held up with my older nephews. It was an out-of-the-box smash hit! They were immediately engaged in figuring out how the parts fit together, before even loading up the build tutorial. It was by far the most popular Christmas toy, keeping five kids (age 3–18) distracted, happy, and quiet for hours. Everyone was so enthralled that the only fight all evening was over which stickers to decorate with.

LittleBits magnetic connectors are easy enough for younger kids, while the color coding is straightforward for older kids. The acrylic droid body is sturdy, and holds up to hard hugs from a 4-year-old. The kids loved the included missions, digging deep into the intuitive Droid Inventor Training App to learn how to control and manage their droid, though there were a few dropouts related to poor cell coverage. As with all littleBits kits, everything is interchangeable and you can keep adding more parts from the ecosystem to extend your droid's abilities. —*Hep Svadja*

INFENTO INVENTOR KIT

$400 infentorides.com

Modular toys like Lego are great, but when you're done putting them together they usually end up on a shelf or in pieces in a bin. The Infento kit is different — it lets you build a number of different wheeled vehicles that you can use as everyday rides. The Inventor Kit has options for eight different builds, from a basic push scooter to a sleek three-wheeled trike with a chain drive and disc brakes. The aluminum extrusion frame pieces are a little blocky but give the ability to quickly reassemble everything on a whim. —*Mike Senese*

LITTLEARM 2C

$99 littlearmrobot.com

The LittleArm 2C is a neat and tidy robot arm kit. The box comes with everything you need, including the Arduino, servos, and plastic bits. Simple instructions make assembly a snap — it took me less than an hour to build mine.

Don't expect world-class strength or accuracy. The tiny servos on this model struggle at times to move with any fluidity or precision, much less lift anything weighing a decent amount. However, this really isn't marketed as a powerhouse or precision instrument either, so that should be no surprise.

The included app allows for remote control over Bluetooth. You can even record and play back actions. Overall, a great kit for experience, but not something you'll put to work. *—Caleb Kraft*

UGEARS HURDY-GURDY

$70 ugears.us

A hurdy-gurdy is a stringed instrument that is played by hand-cranking a wheel that rubs against the strings, like the bow on a violin, and pressing keys to choose the notes to be played. UGears' take gives you everything you need to make a functional instrument. It might not sound the best, but hey, it works. If you're interested in how to build things using a laser cutter, this 292-piece kit is a master class of laser cutting design techniques. It's an intensive build, but when that first note squeaks out it's totally worth the time. *—Matt Stultz*

POLYSHER $300 polymaker.com

People spend hours trying to sand, fill, and even out lines in 3D printed parts to mimic the smooth surfaces of injection molding. But now there's an easier and safer way to get those polished, clean prints with the Polymaker Polysher.

The Polysher is a two-part system. First, print in PolySmooth, a special filament that dissolves in isopropyl alcohol similar to how ABS dissolves in acetone. Next, place the finished print in the enclosed Polysher, which uses a nebulizer to spray a fine mist of alcohol and slowly rotates the print to ensure an even coating. The system is fully automated and easy to use. A large front knob sets the spray time in 5-minute intervals, and when the process is finished a set of fans recapture the alcohol vapor and return it to the reservoir for reuse.

Originally launched on Kickstarter, the Polysher and PolySmooth are now widely available — I got my unit from MatterHackers and had great results printing with the filament. I ran my first smoothing job for 20 minutes and watched as the clear-sided chamber filled with mist. There was a slight alcohol smell, but nowhere near as strong as when I spilled a bit on the floor opening the bottle. The print came out smooth and shiny, though there were still some visible layer lines. A second run took care of most of them. If you have to have consistently smooth prints, the Polysher will likely save you hours of handwork and is well worth the investment. *—Matt Stultz*

SHOW & TELL

Dazzling projects from inventive makers like you

Sharing what you've made is half the joy of making. Want to be featured here? Showcase your projects on makershare.com or tag us on Instagram with #makemagazine.

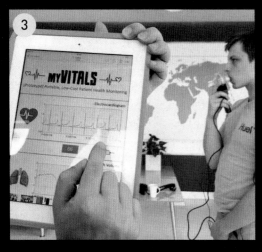

1 **Jonathan Bumstead**'s laser harp produces different notes when the harper passes their hand through the beam. makershare.com/projects/midi-laser-harp

2 This adaptor from **Julio Enrique Rito Vázquez** allows gamers to reach and use every button and joystick on both Joy-Cons with just one hand. makershare.com/projects/single-hand-joy-con-adapter

3 **Mikolaj "Nick" Mroszczak** created myVitals, a low-cost, web-connected, portable medical device, in under a week! makershare.com/projects/myvitals-medical-monitor-information-ag

4 **Hugo Peeters** built this custom console to control his ships in the Kerbal Space Program video game. makezine.com/2018/01/25/kerbal-custom-console

5 These playful slime lamps with special LED strips, created by **The SIProp Team**, can be squeezed and stretched into different shapes. makershare.com/projects/siprop-all-led-project

6 **Ejects Jewelry** combines outdated media with deco-inspired, laser-cut acrylic for a beautiful wearable piece in a retro-futuristic aesthetic. ejectscollection.com